# ART PRACTICE IN A DIGITAL CULTURE

# Digital Research in the Arts and Humanities

*Series Editors*
Marilyn Deegan, Lorna Hughes and Harold Short

Digital technologies are becoming increasingly important to arts and humanities research, expanding the horizons of research methods in all aspects of data capture, investigation, analysis, modelling, presentation and dissemination. This important series will cover a wide range of disciplines with each volume focusing on a particular area, identifying the ways in which technology impacts on specific subjects. The aim is to provide an authoritative reflection of the 'state of the art' in the application of computing and technology. The series will be critical reading for experts in digital humanities and technology issues, and it will also be of wide interest to all scholars working in humanities and arts research.



*Other titles in the series*

Digital Research in the Study of Classical Antiquity
*Edied by Gabriel Bodard and Simon Mahony*
ISBN 9 780 7546 7773 4

Revisualizing Visual Culture
*Edited by Chris Bailey and Hazel Gardiner*
ISBN 978 0 7546 7568 6

# Art Practice in a Digital Culture

*Edited by*

HAZEL GARDINER
*King's College London, UK*

CHARLIE GERE
*Lancaster University, UK*

ASHGATE

Published by
Ashgate Publishing Limited
Wey Court East
Union Road
Farnham
Surrey, GU9 7PT
England

Ashgate Publishing Company
Suite 420
101 Cherry Street
Burlington
VT 05401-4405
USA

www.ashgate.com

**British Library Cataloguing in Publication Data**
Art practice in a digital culture. -- (Digital research in
the arts and humanities)
1. Art and technology. 2. Art and science. 3. Group work in
art. 4. Art--Research. 5. Digital art.
I. Series II. Gardiner, Hazel. III. Gere, Charlie.
701'.05-dc22

**Library of Congress Cataloging-in-Publication Data**
Art practice in a digital culture / [edited] by Hazel Gardiner and Charlie Gere.
p. cm. -- (Digital research in the arts and humanities)
Includes index.
ISBN 978-0-7546-7623-2 (hardback) -- ISBN 978-1-4094-0898-7
(ebook) 1. Art and science. 2. Art and technology. I. Gardiner, Hazel. II. Gere, Charlie.

N72.S3A655 2010
701'.05--dc22

2010008467

ISBN 9780754676232 (hbk)
ISBN 9781409408987 (ebk)

Reprinted 2012

MIX
Paper from responsible sources
FSC www.fsc.org FSC® C018575

Printed and bound in Great Britain by the
MPG Books Group, UK

# Contents

# List of Figures and Tables

**Figures**

**Table**

# List of Plates

# Notes on Contributors

*Paul Brown* is Honorary Visiting Professor of Art and Technology at the Centre for Computational Neuroscience and Robotics (CCNR), University of Sussex in Brighton, UK and Australia Council 'Synapse' artist-in-residence at the Centre for Intelligent Systems Research (CISR), Deakin University in Geelong, Australia. He is an artist and writer who has specialized in art, science and technology since the late 1960s, and in computational and generative art since the mid 1970s. He has participated in shows at major venues such as Tate, the Victoria and Albert Museum and the ICA in the UK; the Adelaide Festival, Australia; ARCO in Spain and the Venice Biennale. He is co-editor of *Leonardo Electronic lmanac* (*LEA*), the e-journal of the International Society for the Arts, Sciences and Technology (MIT Press), and a member of the editorial board of the journal *Digital Creativity* (Routledge). He is Chair of the international Computer Arts Society (CAS) and moderator of the DASH (Digital ArtS Histories) and CAS e-lists. During 2000/2001 he was a New Media Arts Fellow of the Australia Council and spent 2000 as artist-in-residence at the CCNR. From 2002–2005 he was a Visiting Fellow in the School of History of Art, Film and Visual Media at Birkbeck, University of London, where he co-directed the CACHe (Computer Arts, Contexts, Histories, etc. …) project.

*Wayne Clements* is a visual artist and writer. His artworks are exhibited internationally in festivals and exhibitions of electronic art. He programs these artworks using the computer language Perl. He received the Award of Distinction for Net Vision, Prix Ars Electronica 2006. Wayne completed a practice-based PhD in Fine Art, at Chelsea College of Art and Design (2005). His website is www.in-vacua.com.

*Hazel Gardiner* was Senior Project Officer for the AHRC ICT Methods Network (2005–2008), based at the Centre for Computing in the Humanities (CCH) at King's College London. She is joint-editor of the CHArt (the Computers and the History of Art group) Yearbook and a member of the CHArt Committee. She is Editor for the British Academy research project, the Corpus of Romanesque Sculpture in Britain and Ireland (CRSBI), and a researcher for this project and the Corpus Vitrearum Medii Aevi (CVMA), another British Academy research project.

*Charlie Gere* is Head of Department and Reader in New Media Research in the Department of Media, Film, and Cultural Studies, Lancaster University. He is Chair of the Computers and the History of Art group (CHArt), and was the Director of Computer Arts, Contexts, Histories, etc. ... (CACHe), a three-year research project looking at the history of early British computer art. He is the author of *Digital Culture* (Reaktion Books, 2002), *Art, Time and Technology* (Berg, 2006), *Non-Relational Aesthetics: Transmission, the Rules of Engagement,* 13 (Artwords, 2008), with Michael Corris, and co-editor of *White Heat Cold Logic* (MIT Press, 2008), as well as many papers on questions of technology, media and art. In 2007 he co-curated Feedback, a major exhibition on art responsive to instructions, input, or its environment, in Gijon, Northern Spain.

*Beryl Graham* is Professor of New Media Art at the School of Art, Design and Media, University of Sunderland, and co-editor of *CRUMB* (Curatorial Resource for Upstart Media Bliss). She is a writer, curator and educator with many years of professional experience as a media arts organizer, and was head of the photography department at Projects UK, Newcastle, for six years. She curated the international exhibition, Serious Games, for the Laing and Barbican art galleries, and has also worked with the Exploratorium, San Francisco, and San Francisco Camerawork. Her co-authored book *Rethinking Curating* is published in 2010 (MIT Press), her solo-authored *Digital Media Art* was published by Heinemann in 2003, and she has chapters in *The Photographic Image in Digital Culture* (Routledge, 1995) and *Fractal Dreams: New Media in Social Context* (Lawrence and Wishart, 1996). She has presented papers at conferences including Navigating Intelligence (Banff), Museums and the Web (Seattle), and Caught in the Act (Tate Liverpool). Her PhD concerned audience relationships with interactive art in gallery settings, and she has written widely on the subject for books and periodicals including *Leonardo*, *Convergence*, and *Switch*.

*Phil Husbands* is Professor of Artificial Intelligence and Co-Director of the Centre for Computational Neuroscience and Robotics (CCNR), University of Sussex. His research interests include biologically inspired adaptive robotics, evolutionary systems, computational neuroscience, history and philosophy of AI and creative systems. He has a PhD in Computer-aided Engineering from Edinburgh University. He has conducted and led research in bio-inspired adaptive systems and biological modelling since 1985, initially at Edinburgh University before moving to Sussex in 1989. In the early 1990s he co-founded the field of evolutionary robotics. The interdisciplinary ethos of Sussex has allowed him to work closely with neuroscientists for many years, as well as regularly collaborating with philosophers, artists, historians and musicians. He is on the editorial board of eight international journals and has chaired a number of major international conferences.

*Janis Jefferies* is Professor of Visual Arts at Goldsmiths, University of London, UK. She is an artist, writer and curator. She is director of the Constance Howard Resource and Research Centre in Textiles and artistic director of Goldsmiths Digital Studios, an interdisciplinary research centre across art, technology and cultural process. She was one of the founding editors of *Textile: The Journal of Cloth and*

*Culture* in 2002 (Berg) and edited *Digital Dialogues: Textiles and Technology,* 1 and 2 (November 2004, January 2005); a collection of specially-researched and commissioned essays that represent research collaborations between textile artists and designers, cultural theorists, sociologists, architects, computer scientists and engineers. Recent publications include, *Interfaces of Performance*, with co-editors Maria Chatzichristoudoulou and Rachel Zerihan (Ashgate, 2009) and 'Touch Technologies and Museum Access', in *Touch in Museum: Policy and Practice in Object Handling* edited by Helen Chatterjee (Berg Publishers, 2008).

*Gordana Novakovic* belongs to the generation of artists who pioneered electronic art. Originally a painter, with 12 solo exhibitions to her credit, she now has more than 20 years experience of developing and exhibiting large-scale, time-based media projects. Characteristic of her work with new technologies is her distinctive method of creating an effective cross-disciplinary framework for the emergence of synergy through collaboration. She has presented her work at major international festivals and venues, including ISEA, Ars Electronica, ICC (Inter Communication Center – Tokyo), and Tate Modern. Her most recent conference appearances were at Subtle Technologies 2009, and as a keynote speaker at EVA London 2009. Her latest piece, *Fugue* (<http://www.fugueart.com>), has been widely presented and exhibited, most recently at the 'Infectious' group show in the Science Gallery, Dublin, after which it was featured in the October issue of *Nature Immunology*. Since 2004, Gordana has been artist-in-residence at the Computer Science Department, University College London, where in 2005 she founded the Tesla Art and Science Group with colleagues in the department. Tesla is an art and science discussion forum dealing with visionary ideas beyond the existing remits of art and science; it aims to form and nurture cross-disciplinary teams, projects, and networks. Gordana's current work on neuroplastic art explores the possibilities lying at the intersection of art and brain science.

*Jane Prophet* is a British visual artist who works across disciplines with a range of collaborators. She is Professor of Art and Interdisciplinary Computing at Goldsmiths College. Works in development include *Net Work*, a large floating installation (comprising hundreds of illuminated buoys). Her work includes large-scale installations, digital prints and objects such as *(Trans)Plant*, a collapsing and self-assembling sculpture based on the structure of giant hog-weed. Her art reflects her interests in science, technology and landscape. Among her past projects is the award-winning website, *TechnoSphere*, inspired by complexity theory, landscape and artificial life. In 2005 she won a National Endowment for Science, Technology and the Arts Fellowship to develop interdisciplinary artworks. Prophet works on a number of internationally-acclaimed projects that have broken new ground in art, technology and science. In CELL, which began in 2002 she collaborates with Mark d'Inverno, a mathematician, and Neil Theise, a scientist whose ground-breaking research into stem cells and cell behaviour is changing the way that we understand the body.

*Stephen Scrivener* is currently Director of Doctoral Programmes at the CCW Graduate School, University of the Arts, London. He practiced Fine Art during the 1970s, exploring systems-based generative art processes using the processing power of the computer. He later undertook a PhD in computing, and conducted research into computer systems for artists and designers. He also taught computer science until 1999 when he moved back into art and design to further develop his interest in art and design research. During this period he gained experience of a wide range of research methods and practices, which has since served as a ground for his thinking and writing on the theory and practice of practice-based research. He has published a series of papers on this topic, exploring the relationship between art and design processes and products and the conditions of research. His recent writing has also addressed the institutional aspects of this mode of research.

*Paul Sermon* is Professor of Creative Technology at the Research Centre for Art and Design at the University of Salford where he has been based since June 2000. He was awarded the Prix Ars Electronica 'Golden Nica', in the category of interactive art, for the hyper-media installation *Think About the People Now*, in Linz, Austria, September 1991. He produced the ISDN videoconference installation *Telematic Vision* as an Artist in Residence at the Center for Art and Media Technology (ZKM) in Karlsruhe, Germany, from February to November 1993. He received the Sparkey Award from the Interactive Media Festival in Los Angeles, for the telepresent video installation *Telematic Dreaming*, June 1994. From 1993 to 1999 he was employed as Dozent for Media Art at the HGB Academy of Visual Arts in Leipzig, Germany. During this time he continued to produce further interactive telematic installations including *Telamatic Encounter*, in 1996 and *The Tables Turned*, in 1997 for the Ars Electronica Centre in Linz, and the ZKM Media Museum in Karlsruhe. From 1997 to 2001 he was Guest Professor for Performance and Environment at the University of Art and Industrial Design in Linz, Austria.

*Elaine Shemilt* is Professor of Fine Art Printmaking and Dean of External Relations at Duncan of Jordanstone College of Art and Design, Dundee, Scotland. She established Printmaking within the School of Fine Art in 1988, and was its Course Director until 2001. She is a professional member of the Society of Scottish Artists and President from 2007 until 2010. She was elected a Fellow of the Royal Society of Arts in 2000. In 2002 she became a Shackleton Scholar. Her research and practice is diverse in terms of subject, media and content. Her work has often centred on the body. Recent research involves environmental protection in remote environments and genetic coding relating to image pattern and sound sequencing. Outcomes have been produced using etching and screenprinting, video, 3D imaging and animation. These works are exhibited as prints and large scale high-definition projections.

*Stelarc* is an Australian artist who has used prosthetics, robotics, VR systems, the Internet and biotechnology to explore alternate, intimate and involuntary interfaces with the body. His earlier work includes making three films of the inside of his body, amplifying body signals, and twenty-five body suspensions with hooks into the skin. His projects include the *Third Hand*, *Virtual Arm*, *Stomach Sculpture*, *Exoskeleton*, *Extended Arm*, *Prosthetic Head*, *Muscle Machine*, *Partial Head* and *Walking Head*. He is surgically constructing and stem cell growing an *Extra Ear* on his arm that will be Internet-enabled, making it a publicly accessible acoustical organ. In 1997 he was appointed Honorary Professor of Art and Robotics at Carnegie Mellon University. In 2003 he was awarded an Honorary Doctorate by Monash University. He is currently Chair in Performance Art at Brunel University, West London and is Senior Research Fellow in the MARCS Auditory Labs at the University of Western Sydney. His artwork is represented by the Scott Livesey Galleries in Melbourne. The link to his Second Life site is: <http://tr.im/jFGN>.

*Nina Wakeford* is Reader in Sociology at Goldsmiths, and currently holds an ESRC Fellowship that looks at the translations between social science and art/design, particularly in relation to new technology. In 2001 she set up the research group INCITE (Incubator for Critical Inquiry into Technology and Ethnography) (<http://www.studioincite.com>), a creative interdisciplinary space which aims to incubate collaborations between social scientists, designers, engineers and artists. The research group has undertaken several studies of new technologies in collaboration with industrial partners, including a study funded by Intel on ubiquitous computing, and another on new media artists as cultural intermediaries. In her art practice she creates installations, often incorporating time-based elements, which have been shown in gallery spaces and which have also been commissioned to accompany academic conferences.

# Series Preface

*Art Practice in a Digital Culture* is volume eight of *Digital Research in the Arts and Humanities*.

Each of the titles in this series comprises a critical examination of the application of advanced ICT methods in the arts and humanities. That is, the application of formal computationally based methods, in discrete but often interlinked areas of arts and humanities research. Usually developed from Expert Seminars, one of the key activities supported by the Methods Network, these volumes focus on the impact of new technologies in academic research and address issues of fundamental importance to researchers employing advanced methods.

Although generally concerned with particular discipline areas, tools or methods, each title in the series is intended to be broadly accessible to the arts and humanities community as a whole. Individual volumes not only stand alone as guides but collectively form a suite of textbooks reflecting the 'state of the art' in the application of advanced ICT methods within and across arts and humanities disciplines. Each is an important statement of current research at the time of publication, an authoritative voice in the field of digital arts and humanities scholarship.

These publications are the legacy of the AHRC ICT Methods Network and will serve to promote and support the ongoing and increasing recognition of the impact on and vital significance to research of advanced arts and humanities computing methods. The volumes will provide clear evidence of the value of such methods, illustrate methodologies of use and highlight current communities of practice.

Marilyn Deegan, Lorna Hughes, Harold Short<br>
Series Editors

AHRC ICT Methods Network<br>
Centre for Computing in the Humanities<br>
King's College London<br>
2010

**About the AHRC ICT Methods Network**

The aims of the AHRC ICT Methods Network were to promote, support and develop the use of advanced ICT methods in arts and humanities research and to support the cross-disciplinary network of practitioners from institutions around the UK. It was a multi-disciplinary partnership providing a national forum for the exchange and dissemination of expertise in the use of ICT for arts and humanities research. The Methods Network was funded under the AHRC ICT Programme from 2005 to 2008.

The Methods Network Administrative Centre was based at the Centre for Computing in the Humanities (CCH), King's College London. It coordinated and supported all Methods Network activities and publications, as well as developing outreach to, and collaboration with, other centres of excellence in the UK The Methods Network was co-directed by Harold Short, Director of CCH, and Marilyn Deegan, Director of Research Development, at CCH, in partnership with Associate Directors: Mark Greengrass, University of Sheffield; Sandra Kemp, Royal College of Art; Andrew Wathey, Royal Holloway, University of London; Sheila Anderson, Arts and Humanities Data Service (AHDS) (2006–2008); and Tony McEnery, University of Lancaster (2005–2006).

The project website (<http://www.methodsnetwork.ac.uk>) provides access to all Methods Network materials and outputs. In the final year of the project a community site, 'Digital Arts and Humanities' (<http://www.arts-humanities.net>) was initiated as a means to sustain community building and outreach in the field of digital arts and humanities scholarship beyond the Methods Network's funding period.

# Acknowledgements

A number of the authors represented here gave papers at 'Visions and Imagination: Advanced ICT in Art and Science', an event organized by Gordana Novakovic at the Department of Computer Science, University College London, on 24 November 2007, funded by the AHRC ICT Methods Network. This symposium explored methodological questions and problems within the field. The editors are very grateful to Gordana for organizing this event, and, in doing so, making an important contribution to the shape, contents and direction of this book.

# Chapter 1
# Research as Art

Charlie Gere

C.P. Snow has a lot to answer for. A phrase he coined in 1959 for a rather polemic lecture about the division between the humanities and sciences has become an enduring cliché, to be wheeled out every time this supposed division is discussed. As recently as August 2007, Johnjoe McFadden, a professor of molecular genetics at the University of Surrey, and author of a book entitled *Quantum Evolution*, had an article published in the leader section of *The Guardian* newspaper, which starts as follows.

> In his famous Two Cultures lecture, CP Snow lamented the deep divide that separates the arts and humanities in modern culture. But recent work published in Genome Biology by researchers Rie Takahashi and Jeffrey H Miller at the University of California, Los Angeles (UCLA), might be a step towards healing the rift. The scientists designed a computer programme [sic] that turns genes into music. The resulting tunes are surprisingly melodic and have a curious resonance with the roots of both western music and science 26 centuries ago.[1]

McFadden then goes on to describe the program in question which initially involved allocating each of 20 different amino acids a note on the twelve-note chromatic scale (all the notes in the same octave), but finding that the compositions tended to jump sporadically from one note to another and lacked melody, Takahashi and Miller 'reduced the number of possible notes by assigning pairs of similar amino acids to a single note in the seven-note diatonic … scale', and also allowed 'the amino acids to encode three-note chords'. For rhythm, they 'used the frequency of the DNA code that specified each amino acid to assign a time period to each note'. Following this they 'transposed the thymidylate synthase A protein (involved in making DNA) into a pleasant little melody', then a segment of the protein that causes the disease Huntington's chorea, which 'provided a more sombre tune that was interrupted by a repetitive beat denoting a string of glutamines'. McFadden finishes off his article with the following observation.

> It's nearly 50 years since CP Snow delivered his famous lecture, but the arts and sciences are as far apart as ever. Takahashi and Miller's transposition of

1 J. McFadden, 'A Genetic String Band', *The Guardian*, Friday 3 August 2007.

> science into music repays an ancient debt; but perhaps also reminds us that the complementary disciplines have a common root, and once shared the same interests.[2]

Putting aside the rather banal and unimaginative nature of the project itself, what was interesting about McFadden's article was how it revealed an inversion of Snow's point about art and science. Rather than those in the arts and humanities being ignorant of science, the situation is vice versa, with many scientists seeming to know little or nothing about cultural developments and in particular about art and how and why it is practised. It also demonstrated a greater degree of ignorance about historical developments in art in relation to science. It is my experience that, in conferences bringing scientists and artists together, the former are considerably more ignorant about art than the latter are about science. The impression is sometimes given that some scientists at least regard art as being about making pretty pictures, and seem to base their understanding of what, for example, visual art is about on the activities of a small group of then-marginalized French painters working between the 1850s and the 1880s, the Impressionists. Of course, in that it disavows any explicit engagement with politics or culture, and in that it is concerned with the purely optical, Impressionism is obviously congenial to the abstraction and ahistoricity of scientific thinking. This is of course grossly unfair, as there are many scientists who understand and engage with art properly, but I believe not entirely inaccurate. This was the point I made in a letter to *The Guardian* published the next day:

> Johnjoe McFadden's claim that a programme turning gene sequences into music is healing the rift between art and science (Comment, 3 August) ignores decades of collaboration between artists, scientists and engineers that has produced work of considerably more artistic and, more than likely, scientific interest and value.
>
> This rich legacy includes early work in computer graphics and animation by scientists at Bell Labs and elsewhere in the 1960s; the collaborations between artists, scientists and engineers in groups such as Experiments in Art and Technology or the Computer Arts Society in the same period; the investigations into complexity in the 70s and 80s by artists from art schools such as the Slade and scientists in Sante Fe; the long tradition of artistic investigations into the possibilities of robotics and artificial intelligence, from Edward Ihnatowicz to Simon Penny and beyond; and work made in collaboration with scientists about genetics and neuroscience by artists such as Oron Catts, Annie Cattrell, Ruth McLennan and Jane Prophet. Particular mention should be made of the recent work of the Critical Art Ensemble looking at the cultural meaning and effects of biotechnological research – especially given CAE member Steve Kurtz's recent arrest under the Patriot act in the US …

2 Ibid.

> CP Snow's two-cultures argument is wheeled out again as if it is a deep truth about our culture, rather than a now-irrelevant piece of polemic. If there is a rift, it is not in modern culture in general, but in how institutions such as Tate and the Science Museum perpetuate and reinforce unjustifiably hard and fast distinctions between the arts and sciences.[3]

Part of the problem is that there is a kind of cultural cringe at play in the relation between art and science and that we still tend to regard science as the privileged domain in which experimentation takes place, and in which scientific experimental methods are used to find out about the world. Institutions such as science museums tend to reinforce this separation. Yet it can be argued that this presents both a false separation and an overvaluation of science in relation to the rest of culture. Rather than being the master discipline which sets the standards by which we are able to judge truth and knowledge, science is perhaps a particular kind of experimental practice in a more general experimental culture. The experimental is pervasive throughout culture and, for many of us, life is increasingly a kind of experimental process, in which to a lesser or greater extent we have to discover or invent our bodies, ourselves and our communities – culture is the laboratory in which these experiments take place, and our media are some of the principal tools we use. The forms these experiments can take are many and often problematic, including gender reassignment, psychoanalysis, body transformation through plastic surgery, anorexia or bulimia, or prosthetic additions to our biological body, through new kinds of family and other relationships, new kinds of communities, new ways of working, new modes and forms of production.

The roots of this experimental culture can be seen to go back, in the West at least, to the Renaissance in the fifteenth century and the beginnings of the modern scientific world view in the seventeenth century, and to the Romantic ideals of self-creation in the late eighteenth and early nineteenth centuries, as well as the inexorable rise of market capitalism as the dominant force in society. In his famous essay 'Science as Vocation', Max Weber suggests that it is the experimentalism of Renaissance artists such as Leonardo that fosters the scientific method, rather than vice versa.[4] Capitalism, in particular in its current 'late' phase, is predicated on harnessing the experimental drives of its subjects and exploiting the desires these drives make manifest. In the nineteenth century Charles Darwin showed that life is one long experiment, while Karl Marx demanded that we experiment with changing the world, rather than merely describing it, and for Friedrich Nietzsche the death of God required that we engage in a testing process of self-creation. Moving into the twentieth century, Sigmund Freud showed that what we think we know is actually a result of continuing experiments in reality testing. The most

3 C. Gere, 'False Divide Between Art and Science', *The Guardian*, Saturday 4 August 2007.

4 M. Weber, *On Universities: The Power of the State and the Dignity of the Academic Calling* (Chicago: University of Chicago Press, 1976), pp. 141–42.

obvious outcome of experimental culture is science itself, which has transformed our existence dramatically and has also transformed our understanding of the world and our place within it. But science is only one aspect of this culture, and perhaps not the most important.

At the same time experimentalism found expression in the arts with the avant-garde, in which experimentation was both an important strategy and a form of expression – from the early experiments in expression of DADA, the Futurists, the Surrealists, through to postwar avant-garde artists, groups and movements, such as John Cage, Fluxus, early performance, video and conceptual art, as well as those involved in experimental music, free jazz and improvisation. That artists in the mid-twentieth century saw themselves as experimental researchers is nicely indicated by the existence of groups such as *Experiments in Art and Technology*, founded in the United States in 1966, which I mentioned in my *Guardian* letter, or the plethora of so-called 'Arts Labs' that emerged in Britain in the late 1960s. To this might be added the idea of the artist as researcher, or even, in Hal Foster's phrase, artist as ethnographer, that emerged in the 1970s and 1980s.[5] Much of the same spirit of experimentalism, which is mostly lacking in contemporary mainstream art found in galleries such as Tate, is to be found in so-called new media art.

Mainstream art galleries perhaps have trouble recognizing this kind of work as art, at least until sufficient time has passed for it to join the canon, precisely because of its experimental nature. Such work is always, implicitly or explicitly, an experiment about art itself. By extension whatever is produced must exceed what can be recognized as art. If it did not, if it could be easily defined as art, then it would not be experimental. The idea of art as experiment or research must not be mistaken for some kind of 'science envy', in which artists crave some of the institutional respectability of science and its supposedly more secure claims to truth. In a sense the opposite is more true. A great deal of scientific work is not experiment, but testing, in highly defined and restricted circumstances. Unlike the experimental practices undertaken by artists of the sort described above, scientists cannot afford to exceed what can be recognized by their peers as science. Perhaps, only when a Kuhnian 'paradigm shift' takes place does science exceed itself.[6] Paradoxically, artists may be the true experimentalists in culture, rather than scientists, and art is the place where the very question of what exceeds the known can be properly asked. Interestingly, at the edges of what might still be recognized as science, in areas such as artificial life, what is produced resembles (new media) art as much as it resembles scientific research. The word 'experiment' itself can be defined as meaning going beyond the boundaries.

Central to experimental culture is the use of tools, whether they are scientific instruments, information technologies or new media. The more powerful the tools the greater is the capacity to make useful experiments and meaningful statements.

---

5 H. Foster, *The Return of the Real* (Cambridge, MA: MIT Press, 1996).

6 T.S. Kuhn, *The Structure of Scientific Revolutions* (Chicago: University of Chicago Press, 1970).

The greater the degree of access to such tools, the greater is the capacity for experimentation. The kinds of investigations that constitute science, for example, cannot take place without the instruments used. One might go further and say that the 'facts' those investigations supposedly uncover do not exist outside of those instruments. This does not prevent those facts being repeatable and robust, as long as the same kinds of instruments are used, and the facts discovered universally true in that they can be repeated in this manner regardless of the context. But without the apparatus of scientific method, and perhaps most importantly without the observation by scientific peers of experiments, facts of this sort cannot exist for us. Thus, what is produced by science is not a transparent representation of things as they are, but a particular form, albeit both powerful and robust, of contingent enunciation, which produces the truth as much if not more than it represents it (see, for example, Schaffer and Shapin[7] on early modern experimental culture, Latour and Woolgar[8] on 'laboratory life' or Latour[9] on 'science in action').

Perhaps the most important point is that, with the rise of digital or so-called 'new' media, the means of experimentation, production, representation, distribution and consumption are all the same. The technology used by a blogging teen or a member of MySpace, or a net.artist, is more or less the same as that used by a journalist working for a newspaper or, perhaps most importantly, a scientist working on DNA, or artificial life or whatever. This is not to suggest that blogging or being on MySpace or making net.art is the same as those kinds of scientific work, but rather to propose that all are different aspects of a culture in which the experimental is the dominant mode of engagement and production. But there are still essential differences between art and science, and perhaps some are captured by reference to Richard Dawkins's description of scientific method in his essay 'Viruses of the Mind'.

> Scientific ideas ... are subject to a kind of natural selection ... But the selective forces that scrutinize scientific ideas are not arbitrary and capricious. They are exacting, well-honed rules, and they do not favor pointless self-serving behavior. They favor all the virtues laid out in textbooks of standard methodology: testability, evidential support, precision, quantifiability, consistency, intersubjectivity, repeatability, universality, progressiveness, independence of cultural milieu, and so on.[10]

7 S. Shapin and S. Schaffer, *Leviathan and the Air-Pump: Hobbes, Boyle, and the Experimental Life* (Princeton, NJ: Princeton University Press, 1985).

8 B. Latour and S. Woolgar, *Laboratory Life: The Social Construction of Scientific Facts* (Beverly Hills: Sage Publications, 1979).

9 B. Latour, *Science in Action: How to Follow Scientists and Engineers through Society* (Cambridge, MA: Harvard University Press, 1987).

10 R. Dawkins, 'Viruses of the Mind', <http://www.cscs.umich.edu/~crshalizi/Dawkins/viruses-of-the-mind.html> (accessed 15 November 2009).

This then might offer a different understanding of the relation, and difference, between science and art. Science is, perhaps, the name we give to the means by which we apprehend the necessary structure, stricture and coding of the universe, and thus makes it answerable to and describable by precisely those virtues listed by Dawkins: testability, evidential support, precision, quantifiability, consistency, intersubjectivity, repeatability, universality, progressiveness, independence of cultural milieu. These are not criteria imposed upon the universe in some relativistic or constructionist manner but an absolutely integral part of the universe itself, without which there could be no order. Art, on the other hand, is a name for the means by which we come to terms with or remain open to the event, the singular, the monstrous, the contingent, the hazardous, the destinerrant, the elements of play and non-finality within the universe that keep it open to the future, even if this always necessarily involves a process of domestication, much as science must always commence with an openness to play and invention, and an understanding of the necessity of going through the ordeal of the undecidable, that is closer to art than science as we might normally understand it.

In the chapters that follow the relation and differences between art and science in a digital culture are pervasive themes. In Chapter 2, 'Triangulating Artworlds: Gallery, New Media and Academy', Stephen Scrivener and Wayne Clements use the work of James Young and Howard Becker to propose an understanding of the different artworlds in which art is made, and to show how they all both connect and are kept separate, and how they relate to each other. In Chapter 3, 'The Artist as Researcher in a Computer Mediated Culture', Janis Jefferies recounts some of the experiences of artists, including Stanza, Mick Grierson and Tim Hopkins, who have come to Goldsmiths with AHRC Fellowships in the Creative and Performing Arts Scheme, established in 1999 (and no longer running), in the context of a broader discussion of the whole question of the artist as a researcher in higher education, as it has emerged since the 1960s. Jefferies discusses how these artists were able to work with researchers from other disciplines and the effects this had on their practice. Chapter 4, 'A Conversation about Models and Prototypes', is the transcription of an interview between sociologist Nina Wakefield and artist Jane Prophet, known for her collaborations with scientists, about Prophet's work and in particular her interest in and use of rapid prototyping technology. Chapter 5, 'Not Intelligent by Design', by Paul Brown and Phil Husbands, recounts the history of artistic attempts to overcome the effects of the artist's signature style and that of artificial intelligence, followed by a description of the DrawBots project, which aimed to construct a drawing robot, undertaken by Paul and Phil and others at the Cognitive Science department at Sussex University. In Chapter 6, 'Excess and Indifference: Alternate Body Architectures', Stelarc considers some of the questions arising out of his practice as an artist using complex technologies often in the context of universities and other research-intensive organizations. Chapter 7, Gordana Novakovic's 'The Garden of Hybrid Delights: Looking at the Intersection of Art, Science and Technology', recounts her experiences as an artist using technology, both in Yugoslavia and in London, culminating in the Tesla project at University College London. In Chapter 8, 'Limited Edition – Unlimited Image: Can a Science/Art Fusion Move the Boundaries of Visual and Audio Interpretation?', Elaine Shemilt describes her printmaking collaboration with a scientist on the theme of genomics. Chapter 9, Paul Sermon's 'Telematic Practice and Research Discourses', describes three of his telematic projects and finally, Chapter 10, 'Tools, Methods, Practice, Process … and Curation' by Beryl Graham sums up the field of artistic practice in a digital culture, with particular emphasis on the question of curation.

## Chapter 2

# Triangulating Artworlds: Gallery, New Media and Academy

Stephen Scrivener and Wayne Clements

James Young,[1] a philosopher of art, argues for a plurality of artworlds, where understanding of art is coherent within each artworld but not necessarily between artworlds. In this chapter, we use this notion of artworlds, drawing also on the work of Howard Becker,[2] and the practices associated with them, to identify and explore the relations between three artworlds: the gallery artworld, the new media artworld and the academy artworld. There are thus three points, or loci, and these, as we shall see, relate, or fail to relate, to each other in different and complex ways.

Our analysis, we argue here, indicates that there is greater coherence in understanding between research in the academy artworld and the new media artworld. There is also a long-standing and consistent set of relationships between the academic and gallery artworlds. (However, research in the academy may already have begun to trouble these established arrangements.)

We propose that the academic and new media artworlds have deep historic and cultural ties, which, although now somewhat obscured since the institutionalization of research within the art academy, may still provide a basis for continued productive relations between these artworlds in the future. The academic artworld and the gallery artworld are also connected historically and economically. The academy supports artists, for example, and in so doing contributes to the supply of artworks for distribution by the gallery.

Some relations between artworlds are not necessarily harmonious. They are also likely to be subject to change under the impact of new developments. We argue that the new media art world presently poses challenges to the gallery artworld. Its products may circumvent established channels of distribution offered by the gallery artworld, and so evade the curatorial and commercial mechanisms of the gallery and public museum systems. Research within the academy also poses new challenges. Externally, it operates separately from the imperatives of the gallery artworld. Within the academy, staff may be unwilling or unready to view their activity in the context of research. The gallery artworld, correspondingly,

1 J.O. Young, *Art and Knowledge* (London: Routledge, 2001).

2 H.S. Becker, *Artworlds* (Berkeley and Los Angeles, CA: University of California Press, 1982).

does not obviously contribute to, or benefit from, the products of research. It is on this basis that we go so far as to propose that research in the academy is a potential rival system for the development and distribution of artworks, and one in principle favourable to new media artists. We conclude that, with the continued institutionalization of research within the academy artworld, it is possible that the productive relation between new media and research may continue and develop. However, there is no guarantee of this and much depends on understanding the possibilities for productive collaboration between different artworlds based on knowledge of how these artworlds have evolved, and continue to evolve, in relation to each other.

We first develop what we mean by artworlds. We begin by describing the academy, and the gallery and new media artworlds. We then seek to begin to map how our different artworlds relate. We conclude with some potentially difficult implications that art as research poses for the academy and gallery artworlds.

## Artworlds

Becker, exploring the sociology of art, claims that there are artworlds, which ‘consist of all the people whose activities are necessary to the production of the characteristic works which that world, and perhaps others as well, define as art’.[3] He goes on to argue that members of artworlds coordinate the activities by which works are produced, employing conventions embodied in common practices. Hence, an artworld can be understood as a network of cooperative links among participants.

According to Becker, artworlds do not have boundaries around them, separating those who belong to an artworld from those who do not. Instead, Becker argues,

> we look for groups of people who cooperate to produce things that they, at least, call art; having found them, we look for people who are also necessary to that production, gradually building up as complete a picture as we can of the entire cooperating network that radiates out from the work in question. The world exists in the cooperative activity of these people, not as a structure or organization, and we use words like those only as shorthand for the notion of networks of people cooperating.[4]

Here we will focus on identifying three such groups of people: gallery, academy and new media. We will not attempt to map out the entire cooperative network of each of these groups, as advocated by Becker; instead, having identified the artworlds of interest, we will focus on exploring the connections between them so as to reveal the extent to which these connections are mutually supportive or undermining.

---

3 Becker, *Artworlds*, p. 34.

4 Ibid., p. 35.

## The academy artworld

To get to the notion of the academy artworld we need first to consider the traditional relationship between the art academy and the gallery artworld. Historically, the art academy was concerned with teaching. The academy supported the gallery artist in manifold ways, as a source of additional income, a means of influence, a community of like-minded practitioners, a source of refreshment and renewal, *inter alia*, but not as a context for art production and dissemination: art was produced, disseminated and discussed outside the academy and many artists chose to maintain a clear distinction between these worlds.

For many artists this is still the contract. Such artists join the academy to teach and accept the institutional constraints imposed by this responsibility. Their creative work, however, is their own business over which the academy has no rights of control or ownership. However, the contract is changing owing to widespread growth in research in the art academy.

In the UK, the first major stimulant of this growth was the removal, in 1992, of the binary divide which had previously separated the universities from the rest of the higher education sector.

Why should this have been such a powerful stimulant to research? The desire for academic status is probably the short answer. In a survey of UK art and design departments exploring the change factors driving the growth of research in art and design, academic status appeared as the third most frequently occurring change factor and the most frequently occurring positive benefit of change.[5] Motivations identified in the responses for research as a signifier of academic status included: to gain parity with other university disciplines, both those within and beyond a given higher education institution; to gain acceptance for practice-based research (including at doctoral level) as a mode of research; recognition of what other disciplines have achieved through research; a sense of academic inferiority; the need for academic credibility; justification of subject worth; and the socio-cultural prestige associated with research. This is reflected in Christopher Frayling's questioning of whether the debate over the meaning of research, in the context of art and design research, is about '… degrees and validations and academic status, the colour of people's gowns or, more interestingly, a conceptual one, about the bases of what we all do in art and design?'[6]

The second major stimulant to research in the art academy was the UK Research Assessment Exercise (RAE) introduced fully in 1989, after a limited exercise conducted in 1986, and since followed by exercises in 1992, 1996 and

5 S.A.R. Scrivener, 'Change Factors and the Contribution Made by Research to the Disciplines', unpublished report (2007). Scrivener's survey exploring the factors driving the growth of research in art and design was distributed to staff in art and design colleges in the UK. Ninety-six unspoilt questionnaires were returned and analysed.

6 C. Frayling, 'Research in Art and Design', *Royal College of Art Research Papers*, 1/1 (1993/4): 1–5.

2001. The Research Assessment Exercise determines the amount of funding for research that institutions receive from the Higher Education Funding Councils of England, Scotland and Northern Ireland. More or less in parallel with the removal of the binary divide, the non-university higher education sector was first admitted into the exercise in 1992, which apart from academic status brought with it a slice of the research funding cake for which departments of art could compete. The RAE as a source of funding for art research was followed by the establishment in 1998 of the Arts and Humanities Research Board (AHRB), which obtained full research council status in 2005. The availability of sources of funding for art research is the most frequently occurring change factor identified in Scrivener's[7] (2007) survey of the field.

Consequently, in the UK at least, research is being institutionalized as a prominent, integral and defining component of the academy of art: the art academy has become a site for the production of knowledge. This phenomenon is coupled with a debate about the nature of artistic research. On one side of the debate it is argued that research and fine art (i.e. practice) are distinct activities.[8] On the other side, it is argued that art can be understood as research.[9] In between, there are different complexions of research[10] characterized as practice-led or practice-based research, that is, research initiated in practice and carried out through practice.[11]

This plural understanding of artistic research means that the production of art works is now seen as a legitimate function of the academy, thus blurring the boundary between academy and gallery artworld. Whilst problematic, it is not sufficient merely to claim an autonomous academy artworld. For this, we need to show that the production of art in the academy results in a separation between it and the gallery artworld: that is, the academy cannot merely function as part of the gallery artworld network.

Whilst external factors such as the removal of the binary divide and the availability of new sources of external funding have certainly accelerated the growth of research in the academy, it would be wrong to suggest a lack of internal

---

7 Scrivener, 'Change Factors'.

8 Cf. D. Durling, K. Freidman and P. Gutherson, 'Editorial: Debating the Practice-based PhD', *International Journal of Design Sciences and Technology*, 10/2 (2002): 7–18.

9 Cf. K. Macleod and L. Holdridge, 'Introduction', in K. Macleod and L. Holdridge (eds), *Thinking through Art: Reflections on Art as Research* (London: Routledge, 2006), pp. 15–19.

10 S.A.R. Scrivener, 'The Roles of Art and Design Process and Object in Research', in N. Nimkulrat and T. O'Riley (eds). *Reflections and Connections: On the Relationship between Creative Production and Academic Research* [e-book] (Helsinki: Helsinki University of Art and Design, 2009).

11 C. Gray, 'Inquiry through Practice: Developing Appropriate Research Strategies in Art and Design', in P. Strandman (ed.), *No Guru, No Method* (Helsinki: Helsinki University of Arts and Design, 1998), pp. 82–9.

pressure for change. In responses to Scrivener's survey[12] discipline/subject change figured as a major factor in research development. The production of knowledge, i.e. research, is seen as a way of dealing with these internal needs, and hence, whilst art production is seen as central to practice-based research, art production itself is not generally seen as yielding knowledge.[13] In this respect, the academy and gallery artworlds seem to be in agreement: neither sees art as knowledge production.

So while the gallery artworld may accept art produced in practice-based research into its discourse, this discourse cannot be the context for presenting, disseminating and discussing the outcomes of artistic research, since this discourse does not concern art as knowledge. Consequently, artistic researchers must create their own discourse,[14] which, given the institutionalization of artistic research, is likely to be situated and supported within the academy. In this sense it is legitimate to talk about an academy artworld.

## The gallery and new media artworlds

The gallery artworld can be viewed as the normative artworld, supplied by the products of studio practice. It comprises commercial and public galleries, museums and other 'white cube' type spaces where art is entered and re-entered into the wider art discourse. Although Becker distinguishes between private and public spaces, he observes that the 'gallery-dealer system is intimately connected to the institution of the museum. Museums become the final repository of the work which originally enters circulation through dealers'.[15] We therefore consider commercial and public spaces in concert.

There are, within the gallery artworld, considerable uncertainties about what – and how – new media should be managed, conserved, and circulated. This, as we should expect (considering Becker's characterization of the relationship between commercial dealerships and the museum), is the case not only of private dealerships, but also of publicly endowed museums. The following anecdotal evidence[16] represents the extent of this possible polarization.

At a recent conference there was discussion of the preservation of new media art. One of the panellists, an esteemed curator of new media at an important national museum in Scandinavia, spoke about conservation and the need to preserve old

12 Scrivener, 'Change Factors'.

13 However, as we have seen, some scholars are exploring the idea of art as knowledge.

14 The debate surrounding practice-based research can be seen as contributing to the construction of this discourse.

15 Becker, *Artworlds*, p. 117.

16 W. Clements, 'Surveillance and the Art of Software Maintenance: Remarks on logo_wiki', in *Observatori 2008. After the Future* (Valencia: Observatori.com, 2008).

computers and programs and the difficulties of achieving this. Nevertheless, he believed the fundamental project was viable. Someone proposed that an obstacle to preservation was the artist's carelessness, and this view was shared by several of the panellists. Clements replied from the floor that much Internet artwork uses events and content from remote websites over which the artist and the artwork have no control,[17] arguing that this artwork was inherently unstable and temporary. It was intended to be so, and this was part of its unique quality. It cannot be preserved ... The museum curator replied that I had decided (I think I quote accurately) 'to choose to break cultural laws'. We cite this because it illustrates how far it is possible for the new media artworld and the gallery artworld to depart from shared interests.

There are attempts to repair relations. One such is represented by 'Holy Fire' at the iMAL Center for Digital Cultures and Technology, Brussels. This exhibition (18 to 30 April 2008) featured a role call of new media artists. However, several restrictions were issued, aimed, it would seem, at obviating some of the difficulties encountered by the curator quoted above: 'Holy Fire is probably the first exhibition to show only collectable media artworks already on the art market, in the form of traditional media (prints, videos, sculptures) or customized media objects. The artworks collected in Holy Fire are not new media art, but simply art of our time.'[18] Whatever the value set on the claims to novelty, or the reality of the disappearance of new media into 'art of our time', it is clear that new media art continues to pose a challenge to marketing and display in the gallery.

It may be that museums and dealerships will continue to resist works of art that challenge fundamental premises such as durability and stability (and, if truth be told, associated resale value). But we are in a changing situation, and it remains to be seen what the outcome will be of the protean circumstances of emerging conflicts and consensus that comprise the current debates between neighbouring, and sometimes warring, artworlds.

The problems of demarcating artworlds, encountered already, multiply when we address the new media artworld. This devolves in part from the constant revolutionizing of the technological base of new media. What constituted new media when the first practice-based PhD was completed, in 1978, is now no longer new. ICT changes at such a rapid rate that it is constantly outmoded by innovation. There is also the fact of the great diversity of what constitutes 'new media' at any one time. What constitutes the new media artworld is therefore both very diverse *and* constantly changing. This makes conclusions hard to draw (at best). Stephen Wilson's *Information Arts*,[19] a book of nearly one thousand pages, is not at all confident that it is a complete survey of its subject at the time of writing, because this subject is so large.

17 The co-author of this chapter.

18 <http://www.imal.org/HolyFire/en/?page_id=2>. (All URLs current at the time of writing.)

19 S. Wilson, *Information Arts: Intersections of Art, Science, and Technology* (Cambridge, MA: MIT, 2002).

We may add to this that some new media art, comparable to new media communication networks in general, is not composed of discrete objects, but is constituted of a flux of network events. It is this (as remarked above) that makes it difficult or impossible for the dealer or the conservator to isolate and preserve some new media artwork. This network character is much less a potential problem for the academy's research, which is able to set a premium on the challenging and experimental.

Nevertheless, the new media artworld experiences forces emanating from the more established gallery artworlds and academic artworlds which have the potential to influence and reconfigure their activities. An example would be an established new media artist with an art dealer, the latter who seeks to divert the artist into a more conventional career as a video artist and away from more experimental new media artwork.[20]

There is no impermeable barrier between new media art and the gallery artworld. The gallery artworld readily accommodates some forms of new media practice. There are, it is apparent, tensions where the gallery artworld struggles to accommodate fugitive, unstable and difficult-to-conserve art forms. Research in the academy, conversely, is less challenged by such qualities because its cultural norms and activities are different and, we suggest, potentially more conducive to new media.

There are notable attempts by the gallery artworld to include emerging new media within its ambit. An example is the Arnolfini Gallery's 'project.arnolfini', an online venture into new media curation. This includes documentation (texts, images, video) as well as fully curated projects, all accessible through a web browser. As their website states: 'Much of this is in the early stages of development.'[21]

Not only is it still early days, but it also remains a relatively isolated venture. It is hard to predict the outcome of such initiatives. Nevertheless, it is possible to register not only the current obstacles, and attempts to overcome them, in the way of smooth communications between some parts of our triangulation, but also the greater harmony between others. We will also see below that the new media artworld and research in the academy artworld have historic alliances upon which they may draw so as to continue to develop constructive cooperation in the future.

## Relating gallery and academy artworlds

How is the growth and institutionalization of artistic research testing the academy's relationship to the gallery artworld, and other artworlds for that matter?

---

20 Based on a secondhand account of an actual incident, therefore anonymized.

21 <http://project.arnolfini.org.uk/>.

To explore this question we need to consider how the relationship between knowledge and art is being articulated within the discourse on practice-based research and the wider artistic research network: for example, bodies such as the Arts and Humanities Research Council (AHRC) and the Higher Education Funding Councils.

Viewed in the abstract, we can discern four fundamental perspectives on artistic research. We might hold that art is not knowledge and, therefore, that art practices and products have no role to play in artistic research. Alternatively, accepting that art is not knowledge in itself, we might want to claim that artistic practices can contribute to the acquisition and communication of knowledge, while holding that art does not arise as a consequence. Or, being interested in the production of art, we might hold that art can and does arise out of this form of knowledge production: in other words, we might want to claim both the production of knowledge and art. Finally, we might simply claim that art is knowledge. All of these perspectives can be discerned within the scholarly debate on practice-based research. Each can be seen as having different implications for the relationship between the academy and the gallery artworld.

Earlier it was claimed that it was appropriate to talk about an academy artworld, because by embracing practice-based research as an institutional goal, the art academy has also committed itself to the production, dissemination and discussion of art. We can now see that this only applies to the third and fourth perspectives, because only these claim both knowledge and art.

Turning to consider the potential implications of these different perspectives, the first is perhaps the least problematic at the institutional level, since art and knowledge discourses are separate. Nevertheless, the growth of this view of artistic research will have a direct and indirect impact on the gallery artworld through the redirection of an academic resource that has traditionally supported artistic production. Contractually, the academic is typically required to engage in scholarship, research and/or professional practice (normally 20 per cent of contracted time). Historically, it has been in the art academy's interest to allow the practitioner to use this time to maintain his or her practice.[22] With the growth of research, the art academy is redirecting effort toward research, which, from the perspective under consideration here, means that art making must give way to knowledge production. If this change results in a reduction in the quality and quantity of the art supplied to the gallery artworld, we should expect inter-institutional tensions to arise. For the individual gallery artist working in this context, it is not simply a matter of having less time to engage in practice; the artist must now acquire a new commitment to and competence in artistic research in the absence of a compelling reason for doing so, other than institutional imperative.

Although the above points apply in relation to the second and third perspectives, we might be forgiven for thinking that the effects are likely to be less damaging, since artistic practices are engaged and art is produced. However, each in fact

22 This is further supported by the lack of institutional activity during vacation periods.

brings with it additional complications. In the case of the second perspective, since research involves artistic practices, we can expect that both the gallery artworld and artist will have anxieties about the potentially damaging consequences of such appropriation.

From the third perspective, both art practices and art itself are appropriated. We have already observed that the gallery artworld discourse cannot be the discourse for practice-based research because it is not a knowledge discourse. This being the case, a further potential point of conflict exists in that art must now function in two discourses – academic and gallery.[23] Institutional conflict is to be expected since art has both cultural and fiscal value. Having invested in the production of that capital, the art academy will want to benefit from it: indeed, such benefit provides a justification for the investment in the first place. We should therefore expect the academy artworld to develop means of presenting and disseminating the art products of practice-based research. Unless developed collaboratively, these means are bound to be in conflict with the gallery artworld. Individual artists working in an art academy adhering to the 'knowledge and art' perspective on practice-based research will be presented with new challenges in managing the demands of the two artworlds on their time and creative outputs.

The above also applies to the fourth perspective, art as knowledge, but here the object of the academy and gallery artworld discourses is the same, i.e. art – the former adhering to a less catholic definition of art than that accommodated by the latter. From this perspective, there is little scope for the artist to maintain a separation between the art academy and art practice, since the latter, viewed as research, serves institutional purpose. In this context, to be an academic now means that the artist's creative work is no longer entirely their own business and therefore a business over which the academy has rights of control or ownership.

This discussion of the relationship between the academy and gallery artworlds reveals change that presents many challenges for both institutions and the artists operating between them. In fact, the situation is even more complex than that suggested by the above. Indeed, you would be hard pressed to find any art college that adheres to a single perspective. Most perspectives appear to be circulating in most institutions.

The situation is confused and confusing,[24] as reflected in the results of Scrivener's survey,[25] where conflict between research and art/design practice emerged as the most frequently reported negative consequence of research growth. Table 2.1 shows some of the conflicts identified by respondents that can be seen as consistent with the situation characterized above.

---

23 This would not be the case if the knowledge produced can enter into the discourse unaccompanied by the art.

24 It is worth noting that the RAE definition of research can be viewed as admitting advanced practice as research.

25 Scrivener, 'Change Factors'.

**Table 2.1 A sample of responses categorized as indicating conflict between research and art/design practice**

| |
|---|
| Good research might produce bad art |
| Research overemphasized at the expense of creative production |
| Misunderstanding surrounding practice and research |
| A move away from art, assumed as 'pure' research by creative practitioners |
| Tendency to interpret 'field of inquiry' as the same thing as 'field of practice' |
| Uncomfortable relation between research in the context of art and art in the context of research |
| Scepticism within art world(s) over the purpose of artists doing research |
| Largely external agendas are seen as beginning to lead practice |
| Projects/creativity being shaped by research |
| Devaluing of research, to make it more accessible to practitioners |
| Precedence of the text over the artwork |
| Diminished private practice e.g., galleries uninterested in research outcomes |
| Undermining of the academic status of practice |
| The view that research is about academe, not the artworld, suggesting a divorce between the art world and art within the academy |

*Source*: Adapted from Scrivener, 2007.

Table 2.1 reveals not only the fact of conflicts, but also that their expression reflects different perspectives on the situation; perspectives which are easily connected with one or other of the logical perspectives described above.

## The gallery practitioner and the radical practitioner

The discussion paints a conflictual view of the relations between the academy (research) and gallery artworlds. Put another way, although these artworlds have an interest in the art, the nature of their individual interests seems to be conflicted, but why should this be the case?

In the first place, the academy artworld is being shaped largely by external imperatives. Furthermore, the gallery artworld does not appear to be one of these forces. In fact, the gallery artworld appears, in the main, uninterested in artistic research. Instead, the desire for academic status and better resources are more readily seen as major drivers of change, and little evidence can be found to suggest that research development reflects academic need to rethink the nature and purpose of art.

Conflict in itself does not point to insurmountable institutional conflict since, even in institutional cooperation, conflict is likely to arise. However, effective cooperation requires clarity of purposes and goals and a clear understanding of how these will be advanced through cooperation. The above discussion points to a lack of clarity about the role and purpose of artistic research, particularly practice-based research. For the present at least, whilst the academy and gallery artworlds appear to have a common interest in art, the institutions are not in a position to know whether they should compete, cooperate or simply ignore each other. Rather than being in conflict, it is more accurate to suggest that these institutions lack a clear understanding of how they relate to each other under the emergence of an academy artworld. In practice, this uncertainty is registered as conflict by individuals working across the boundary between strongly institutionalized artworlds, particularly those artists highly embedded in the gallery artworld.

Of course, there are many artworlds – the gallery artworld is just one of these. In fact, it can be argued that even the gallery artworld, as we have described it, encompasses a number of artworlds, which hold to uncontroversial definitions of art. Following the emergence of the idea of the avant-garde in the earlier twentieth century, a wide range of experimental and radical definitions of art have been explored and continue to be explored, notwithstanding the demise of the avant-garde. For the purposes of this discussion, we shall call these 'radical artworlds'. Importantly, these radical artworlds tend to emerge out of the desire for change amongst like-minded individuals in reaction to the gallery artworld. Furthermore, they tend to be relatively small in scale, informally constituted, suspicious of institutions and ideologies, and yet able to move across a range of contexts.

Unlike gallery practitioners, radical practitioners are not closely connected to the institutionalized artworld, that is, the gallery artworld. Traditionally, the fine art academy has provided a haven for avant-garde and experimental art practices, exploring definitions of art and producing outcomes at odds with the dominant gallery artworld. Consequently, radical practitioners tend to operate outside the demands of the gallery artworld and yet subsist on its practices and resources to some degree or another. Nevertheless, like gallery practitioners, they have maintained a distance between their art practice and the art academy.

Given this lack of attachment to the gallery artworld, there is an argument for suggesting that the radical practitioner will not experience the same tensions between contexts as that experienced by the gallery practitioner. Nevertheless, the radical practitioner is likely to be resistant to the emergence of an academy artworld both conceptually and practically, in so far as the purposes and practices of the institution conflict with their purpose and practices.

Whilst the above discussion has largely focused on the problematic aspects of the growth of research in the academy, in the remaining sections we argue that the history and development of the new media artworld reveal practices that are highly consonant with the research practices of other academic subjects. Indeed, we will argue that the new media artworld has utilized the culture and resources of the academic research world for its own purposes. As such, there is a case for suggesting that the new media artworld has a positive role to play in the development of the academy artworld by providing models for the development of new art discourses.

## Relating new media, academy and gallery artworlds

We have suggested that there is a possible basis for radical artists seeking support in the academy, rather than in the gallery artworld. We suggested that this is because it may act as a haven for artists who are not necessarily so well integrated into the gallery artworld. But there might also be other positive bases for the convergence of art as research and new media. The orientation of new media art to scientific and technological developments *may* make it more sympathetic to the idea of research *per se*. There is, in short, a shared culture of acceptance of research as an activity between new media artists and the world of information and communication technology. This is not to go so far as Wilson, who suggests that:

> Perhaps the segmented categorization of artist and researcher will itself prove to be an historical anachronism; perhaps new kinds of integrated roles will develop … hackers who pioneered microcomputer developments may one day be seen as artists because of their intensity and culturally revolutionary views.[26]

It is not necessary to propose that technologists are artists or that artists need make technological innovations to merit the title of researcher. Rather it is that research is a mutually acceptable activity and a way of defining the activities of both new media artists and new media technologists.

Thus, we can argue that there is greater coherence in understanding between academic and new media artworlds based on a shared acceptance of research. We also argue that it is matched by less coherence between the *research* academic artworld and the gallery artworld. We say specifically 'the *research* academic artworld', as research is still not wholly accepted, either as a possible academic pursuit, or, if possible, not as a welcome or helpful one, and is one, in fact, that faces criticism and even opposition in some parts of the art college (as touched upon above).

There are tensions not only between (or inter) artworlds, but also within (or intra) artworlds. The gallery artworld, as we have seen, cannot be typified as unified and homogeneous. Here too we can find initiatives to encourage and accommodate new media art practice. Research also continues to be resisted by parts of the academic artworld. This is particularly so in those sections of the fine art academy which have little traditional base in art as research. The situation, in short, is complex.

The new media artworld has no historical or philosophical resistance to the idea of art as research. Therefore, forms of practice, including new media practice, have been able effectively to lead research in fine art and have done so from its earliest days. The gallery artworld and traditional media departments historically have been much slower to come on board. This reluctance is both old and deep-rooted. So, in 1983 at the seminar 'Fine Art Studies in Higher Education' at

---

26 Wilson, *Information Arts*, p. 50.

Leicester Polytechnic, called to discuss the question of research in fine art, doubts were raised about the necessity of the whole enterprise. It was demanded: 'Is there any need for fundamental research when we know instinctively the nature and value of practices such as drawing?'[27] (Such viewpoints, it may be suspected, continue to influence contemporary attitudes in some quarters of the art college.) Paradoxically, as we will see shortly, institutions such as Leicester Polytechnic were in the forefront of developments in new media and in art as research.

Becker discusses how artworks require collective activity for their existence. Much of this activity is hidden from view, such as, for instance, the creation of artists' materials and the making of musicians' instruments. Artworlds also have different tasks in the production and public circulation of artworks. Distributing artworks is one central activity for any form of art. The issue of distribution of artworks is closely linked to the need to answer the question of how to 'repay the investment of time, money, and materials in the work so that more time, money, and cooperative activity will be available with which to make more works'.[28] The modern commercial gallery has an important role in the distribution of artworks. At its most fundamental level this is achieved through the mechanism of sale of work. It is not, of course, the only method of supporting art and artists. Others include self-support, important for many forms of challenging or experimental work, and, historically, patronage, although commercial sales of one form or another have largely replaced this.

New media artworlds share with the research context in art colleges and art departments in universities the fact that the requirement to find an immediate return on their activities through commercial exchange may be deferred – perhaps indefinitely, but at any rate for a considerable time. Moreover, the research context's relation to that of material return is highly mediated and not predicated on an expectation of direct exchange relations. Rather its goals are longer term and based on compensation for exploratory and development work that may find no immediate financial recompense. It does, of course, aim to raise status and this *is* ultimately related to funding. The contention here is that the new media artworld and the research artworld share some common values and practices that are more mutually compatible than they are compatible with the practices and values of the gallery artworld. This compatibility–incompatibility relationship is not absolute, but rather should be characterized as tendencies within and between artworlds. These tendencies form the basis of an historical alliance between the new media artworld and the research academic artworld. This relationship is not exclusive, as the alliance includes other forms of challenging art practice, particularly conceptual and feminist art practices. These alliances continue to influence the contemporary

27 C.R. Brighton, 'Research in Fine Art: An Epistemological and Empirical Study', Sussex University, unpublished PhD thesis (1992), p. 316.

28 Becker, *Artworlds*, p. 93.

situation in the United Kingdom.[29] However, there is a good case to be made for the argument that new media art has led the way in fine art research and that it may continue to do so. It is also true that the 'old media artworld', if it may be put like that, struggles to accommodate itself to the new situation of art as research.

The new media artworld and *research* in the academic artworld are in a position to share some compatible understandings of what art may be. This includes the encouragement of interdisciplinary working and freedom from commercial constraint. These elements of shared coherency are not only philosophical and practical but also historical. There is a clear historical alliance between fine art research and research specifically in new media. Significantly, the first PhD to be awarded in fine art was not in a traditional art college but in a polytechnic (Leicester) and was for research undertaken by Andrew Stonyer, in the new field of solar kinetics. Catherine Mason writes:

> Leicester became the first institution in the UK to award the PhD in Fine Art in 1978 to Andrew Stonyer, an artist working in solar kinetics – what today would be termed New Media. Stonyer's supervisors were the Head of Sculpture and a Reader in Chemistry, with the Slade School of Art as the collaborating establishment. Arguably such an accomplishment would not have been possible outside a Polytechnic, which was pioneering in its collaborative research-based culture.[30]

C.R. Brighton, in an analysis of the earliest registrations by the CNAA of research degrees, found, 'some twenty applications which can be classified as 'fine art' including some aspect of practice'.[31] The second research degree to be completed in the United Kingdom was by L.P. Newton, an MPhil in 1981 at Wolverhampton Polytechnic. Its title was 'A Computer Assisted Investigation of Structure, Content and Form in Non-figurative Visual Imagery'. Here again we find a new media artist leading the way in research, again not at a mainstream art college. Of those for which information exists, it is interesting to find that, with one exception, art

29 F. Candlin, 'Artwork and the Boundaries of Academia: A Theoretical/Practical Negotiation of Contemporary Art Practice within the Conventions of Academic Research', University of Keele, unpublished PhD thesis (1998), makes a strong case for the importance of these practices in several academic centres in the UK. These include Leeds Metropolitan University and Keele University, as well as the Royal College of Art (p. 58). The latter is exceptional for an established art college.

30 C. Mason, 'A Computer in the Art Room', <http://www.chart.ac.uk/chart2004/papers/mason.html> (2004).

31 Brighton, *Research in Fine Art*, pp. 351–2. Of these, four were PhD awards, three were MPhils, two were withdrawn, two were rejected, one remained registered for thirteen years, and eight seem unaccounted for (p. 351). Two of these degrees, at least, were investigations of sculptural practices.

departments do not seem to figure in these early ventures.[32] Two other research degrees were undertaken at Newcastle-under-Lyme Polytechnic. (A qualifying point needs to be made that according to Brighton's research, some of these PhDs were in more traditional media, such as sculpture and painting.)

Leeds Metropolitan University and Keele University were also, according to Fiona Candlin,[33] among the first to award practice-based PhDs. The practice parts of both her research and that of the second successful candidate, Allessandro Imperato, were in new media. Imperato investigated surveillance and the 'practice involved digital media'.[34]

From all this it is clear that there was a tendency for research in art to be fostered outside the art colleges and departments, and for it to be undertaken (albeit not exclusively) in new media practice. This legacy has a bearing on the current situation as well as on future prospects for fine art research. From its beginnings in the late 1970s to early 1980s, both research in art and the practice of new media art has changed beyond recognition. There is now, if not a universal acceptance, an institutionalization of both research and new media within the academy, which is now, reluctantly or not, involved in research if only because, as Frayling[35] (2006) has pointed out, funding is increasingly tied to research outcomes. This is mirrored by a partial acceptance of some new media practice within the gallery artworld. However, these accommodations are by no means complete. Photography and video, for instance, may be counted 'new media', and are exhibited, but are well-established practices. In this sense, they can scarcely any longer be regarded as new. But, as we have seen, other forms of new media may provide greater challenges for galleries, dealers and curators.

Research departments support new media art institutionally. It is also the case that the public distribution of this art is assisted by a network of festivals and events, abetted by a relatively small number of specialist art galleries. New media artists are able to work internationally and, at least in part, outside the gallery system, facilitated by the great number of festivals of new media art. These festivals and events do not, as art colleges do not, have an express 'mission to sell'. Rather they rest on the basis of an inscrutable mix of cooperation, voluntarism and public funding.

An example is the SHARE festival, based in Turin: 'an international gathering for digital art and culture'.[36] SHARE exhibits electronic artworks, and awards a prize each year for innovative digital art. It lists as its supporters the City of Turin;

---

32 The exception is Exeter College of Art and Design, where the head of Fine Art, A. Goodwin, completed a PhD in 1982.

33 Candlin, 'Artwork and the Boundaries of Academia', p. 29.

34 *A_Imperato_Resume*, <http://alessandroimperato.com/Resume-Contact_files/A_Imperato_Resume.pdf>.

35 C. Frayling, 'Foreword', in K. MacLeod and L. Holdridge (eds), *Thinking through Art: Reflections on Art as Research* (London: Routledge, 2006), pp. xiii–xiv.

36 <http://www.toshare.it/eng/>.

the philanthropic Compagna di San Paolo, a private non-profit organization; Fondazione CRT, also a private non-profit organization established in 1991; and the Regione Piemonte and the Provincia Di Torino.

The festival permits the presentation of new media artworks, such as the exhibition RADICAL SOFTWARE, 2006, curated by Domenico Quaranta, in which one of the authors of this chapter participated as well as other artists in new media.

The situation is far from stable and is changing rapidly. It is truly impossible to predict what will happen in research in fine art, new media, or in the exhibition and curation of new media art. The latter is perhaps the most unpredictable and it is possible that the gallery artworld may improve its current rather poor response to emerging new media and to research.

## Conclusion

The above discussion poses complex issues. Returning to the section above ('Relating gallery and academy artworlds'), it was found that how research was conceived in relation to knowledge determined the relations of the artwork to the academy. Proprietorial and fiscal consequences were not least among these. The issue of new media adds a complicating aspect to this already complex area. If new media artists, specifically in the form of the radical practitioner characterized above, continue to create art that seeks to evade and even defeat the gallery artworld, some important potential consequences do follow. It is only possible, in short, for the academy to seek control of the products of new media researchers if these products are truly controllable. It will be recalled that of our four perspectives, the second, third and fourth all led to the academy having some claim on the artistic products of research, in so far as in these perspectives artworks constituted knowledge (albeit in different ways and to different degrees). This is because in these perspectives, art either *is* knowledge or it contributes *to* knowledge – *and* the production of knowledge is the business of research in the academy. Thus having funded its production, it has a claim to its products. However, this is only possible, of course, where there is something that can be claimed effectively. We say 'effectively' because although immaterial goods may be the objects of legal limitations, these legalities must be enforceable for them to be a reality. In other words, will the academy be able or willing to catch up with the digital age so as specifically to appropriate products that might be more successfully licensed in some form of open source agreement or similar arrangement? Unfortunately this question lies beyond the scope of the present chapter. Such possibilities have scarcely begun to register in the current debates in the academy artworld.

Whatever the outcome of new initiatives and emerging situations, there is, nevertheless, a challenge posed not merely to the gallery artworld but also to the academic artworld, its research part included. It is possible that one or both of these artworlds will fail to rise to these challenges. The historical alliance between new media and research in fine art does not provide a guarantee that their historic coherence will continue. Its beginnings were quite long ago and are largely unrecognized and forgotten. It may be that artists whose work embraces the ephemeral and immaterial may still gravitate to centres of learning that can offer them support. If so, the relationship is not likely to be one-way. The presence of these artists will provide an internal challenge to any entrenched and traditionalist interests that may persist within these institutions.

## Acknowledgements

The research reported in this paper was supported by the Arts and Humanities Research Council, UK, under grant number 112155.

# Chapter 3
# The Artist as Researcher in a Computer Mediated Culture

Janis Jefferies

The Arts and Humanities Research Council (UK) established their Fellowships in the Creative and Performing Arts Scheme in 1999. The primary intention was to support practice- and performance-based researchers who had not had the opportunity to undertake a significant programme of research within a higher education environment. One of the consequences of the scheme has been to encourage artists to pursue their careers as active academic researchers. This chapter will explore some aspects of the paradigm shift from the artist as practitioner to the artist as researcher, and how the idea of visual arts practice has been transformed in the process. The main objective of this chapter is to define how a new generation of artists and AHRC Creative and Performing Arts Fellows function within a computer-mediated culture, in a higher education environment.

## Pigment, pixel, process

In the recent past, independent art schools in many countries, including the UK, provided discipline-specific courses that mostly drew on the atelier traditions of the academy or on Bauhaus-inspired formalism. This indeed was my own experience at art school both in England and later in Poland. Many of us, particularly women (including myself) who studied painting in art school in the early 1970s, became disillusioned with what we felt to be a narrow definition of art and the need to uphold a restrictive canon. The debate that ensued about whether an artist was 'made' or could be 'taught' owes a great deal to arguments about what constitutes skill, whether it is part of an art education, training or a networking opportunity to enter the professional (and at that time) singular version of the art world, its institutions and processes of commodification. To my knowledge the word 'research' was never used within a studio context but referred to only within complementary studies, developed from art history, with a broader cultural remit. The now defunct Council for National Academic Awards (CNAA) ratified my own degree in Fine Art at Maidstone College of Art in 1974. The following year it was possible to have the degree award of BA(Hons) if one's creative work was clearly presented in relation to the argument of a written thesis. This had to be set in the relevant theoretical, historical or critical context. The restructuring of art and design education along university models was instigated by the first of several Coldstream reports. For example, during the 1960s William Coldstream and his committee made a report that changed the landscape of art education in the UK forever. Did he foresee that the acceleration of technology and what was happening in art history and theory demanded that art schools would never become respectable unless they were able to have degree-awarding powers? During the 1970s critical and contextual studies were established on the newly-formed Honours programmes that were to provide a research base for the new universities.[1] As a consequence, a new era of intellectually ambitious critical self-reflection and cultural renewal was trumpeted. On reflection, and after 30 years working in art schools and universities, I would argue that an unresolved tension emerged between what was constituted as academic and scholarly research and the artist theorist who raided across ideas, disciplines and practices to invigorate visual arts practices from within.

However, and partly due to the integration of art schools and art departments within a UK higher education environment, the full dimension of how visual arts can be seen within a broader 'arts based' set of practices has only recently been acknowledged.[2] As in the 1980s, there is a case to be made for the image of the artist-theorist as practitioner and researcher, rather than simply as arts educator. The practitioner-theorist came to the fore during the 1980s, and I was certainly included in that category, employed both as visiting artist and theorist. It was also an era in which scripto-visual/text and image production dominated debates within the studio and the academy.[3] For example Victor Burgin, in particular, was courted by the academy, exploiting connections between practice and academia at the same time as occupying an uneasy position within artistic production.[4] I would not be the first to point out that Burgin, the former Millard Professor of Fine Art at Goldsmiths, University of London, pursued a practice inseparable from his theoretical writings which are steeped in the ideas of many twentieth-century poetical, psychoanalytical and linguistic theorists. During the 1970s and 1980s his work was based on the juxtaposition of text, and image names, as the 'scripto-visual'. *The End of Art Theory: Criticism and Postmodernity* (1986) was extremely influential in developing cultural criticism. What came to be known as

---

1 For a full account of this debate, see J. Thompson, 'Art Education: From Coldstream to QAA', *Critical Quarterly*, 47/1–2 (2005): 215–55.

2 F. Candlin, *Working Papers in Art and Design*, vol. 1 (2000), <http://www.herts.ac.uk/artdes1/research/papers/wpades/vol1/candlin2full.html>. (All URLs current at the time of writing.)

3 One strand of feminist art theory, sometimes referred to as 'scripto-visual' practice, is close to the semiotically-informed critical practice outlined in Elizabeth Chaplin's, *Sociology and Visual Representation* (London: Routledge, 1994).

4 Victor Burgin was appointed Millard Professor of Fine Art at Goldsmiths, University of London in 2001 and until 2007. I was head of Visual Arts between 2002 and 2004 before I transferred to the department of Computing to establish Goldsmiths Digital Studios.

practice-based and practice-led research (particularly in the visual arts) has to now satisfy both the demands of the university as well as the non-academic structures of art production. This is an area that continues to be contentious and troubling.[5] It is possible that the institutional reception of Burgin's work helped to establish a template for artistic practice and the dominance of text in the university.[6]

Increasingly, the visual arts, as practice-based and practice-led research, are positioned as having to be grounded in practices from art itself, particularly if the enquiry is studio based,[7] although there has been fervent debate around practice-based research in the UK, Northern Europe and Australia since the 1990s. The Creativity and Cognition Studios (CCS), led by Professor Ernest Edmunds (University of Technology, Sydney, Australia) has identified how the terms 'practice-based' and 'practice-led' research are increasingly interchangeable. Within CCS definitions, 'practice-based research' is an original investigation undertaken in order to gain new knowledge partly by means of practice and the outcomes of that practice. Creative outcomes can include artefacts such as images, music, designs, models, digital media, performances and exhibitions. 'Practice-led' research is about practice, and results in new knowledge that has operational significance for that practice. The results of practice-led research may be fully described in text and it is this relationship to writing that leads to contentious debate.[8]

5 For example, Mick Wilson, Dean of the Graduate School of Creative Arts and Media completed his research, 'Conflicted Faculties: Knowledge Conflict and the University' at the National College of Art and Design, Dublin in 2006. It informed the international conference 'Arts Research: The State of Play' organized by the college in May 2008. This conference explored the current state of the field (with particular reference to developments since 2005) in terms of case study examples of doctoral arts-research practice, examples of contemporary arts-research practice beyond the academy, international interaction and networking in arts-research.

6 M. Connolly, *Art Practice, Peer-Review and the Audience for Academic Research*, Position Papers on Practice-Based Research, National College of Art and Design, Dublin, Ireland, 22 April 2005.

7 G. Sullivan, *Art Practice as Research: Inquiry in the Visual Arts* (London: Sage Publications, 2005).

8 L. Candy, 'Practice Based Research: A Guide', Creativity and Cognition Studios, University of Technology, Sydney, CCS Report: 2006-V1.0 November 2006. Available as a PDF via <http://www.creativityandcognition.com>. The author and Linda Candy were in email exchange about the content and production of this document throughout 2006. See also, M. Makela, 'Knowing Through Making: The Role of the Artefact in Practice-led Research', *Journal of Knowledge, Technology and Policy*, 20/3 (2007) available online. In this paper, and mapping the debate in Finland, the key questions are identified as follows: 'The central methodological question of this emerging field of research is: how can art or design practice interact with research in such a manner that they will together produce new knowledge, create a new point of view or form new, creative ways of doing research? That which is produced by an artist-researcher, the artefact, can also be seen as a method for collecting and preserving information and understanding.'

Some of the same issues that beset practice-based and practice-led research within a higher education environment are also faced by some of the AHRC Fellows with whom I have institutional and professional contact. They are entering into the larger academic enterprise for the first time as creative practitioners. One Fellow describes his process as follows:

> Practice and development for me are interchangeable. The research is the pre-development of everything that I do. The practice is the actions that are employed because of the knowledge gained through research. Research is about ideas and concepts and working towards model prototypes. For me research is defined by definitions of my practice and that of all practice-based artists. What I am doing is making a contribution to the field of knowledge or artistic experience that adds to their historical experience.[9]

As described by another AHRC Fellow, the relationship between the definitions of research and method and outcome required by the AHRC does not form a seamless fit with all aspects of the realization of practice-based research:

> The process of reflecting on music and performance (the framework of a traditional academic context) comes with a completely different way of thinking and communicating to making work *per se*. Practice as research appears to challenge this as the application process itself requires detailed structural information, a prediction of outcomes and description of method of approach to an exacting degree. I think this may be a point of tension. I think this may be an issue for other artists engaged for the first time in the academic world, where their practice has developed primarily outside it.[10]

---

9 Email correspondence between the author and Stanza, AHRC Creative and Performing Arts Fellow, Goldsmiths Digital Studios, Goldsmiths, University of London, UK (2006–2009), 17 April and follow-up meeting on 22 April 2008. We normally meet once a month to discuss the progress of his research. Interestingly, Stanza has maths, physics and art A levels (1979); an Art and Art History degree from Goldsmiths College (1985); a PgDip in Visual Communications from Goldsmiths College (1987); and a MA in Multimedia from Central Saint Martins School of Art (1995) and so is familiar with the academy and working across disciplines.

10 Email correspondence between the author and Tim Hopkins, AHRC Fellow in the Creative and Performing Arts, Centre for Research in Opera and Music Theatre, Department of Music, University of Sussex, UK (part-time, 2007–2012), 17 May 2008 and subsequent interview on 28 May. Hopkins is undertaking research relating to a number of projects that are investigating the potential of new technologies and media for lyric theatre. The projects include multimedia performance, interactive installation, and adaptations for the small screen. Initial studio-based research has employed digital tools, building on previous experience with sound/image translation devices MaxMsp/Jitter and Isadora. Goldsmiths Digital Studios supported his application to the AHRC and is a co-partner for seminars and events.

As I know, artists' studios, and other such environments within a university context, can be significant places for the creation and the critique of new knowledge, which is what practice-based research demands at one and the same time – both theoretically powerful and methodologically imaginative. In the twenty-first century the idea of visual arts practice is again being expanded, advancing our understanding of who we are, what we do and what we know. Digital environments, studio/labs and cultural collaborations offer new and different forms of research and scholarship. As pointed out by Graeme Sullivan: 'a range of models of practice evolved as history moved from the café, from the classroom to the studio, and into the virtual world'.[11]

In the next section of this chapter I will discuss how a generation of artists, including myself, pursue diverse models of research within a computer-mediated culture and in an academic environment. How our work unfolds will become a crucial part of our cultural heritage. That this work is often done in partnership with engineers and scientists is reflected by the ways in which some of us work in multidisciplinary teams and on particular aspects of work. Frequently there is cross-collaboration with writers and curators in order to disseminate practice to a wider audience. This, in the future, will be another challenge as the complex forms of dissemination shift from physical space to the Internet and back again. At the same time we are pursuing individual artistic practice and frequently engaged in the development of technologies that are shaping our society. For example, for Stanza and Mick Grierson, research in the development of technologies is informed by the possibilities afforded by a broad range of art practices that affect the ways in which we perceive, process and respond to visual information. Of course, the intersection of art and technology is not new but finding new terminology for emerging art and technological practices can be fraught with difficulty. What we might call computational culture, is drifting and expanding as fast as expanding definitions of art are being challenged: terminals are no longer fixed, art comes at you from many directions and at the speed of light, distributed media pervade our everyday existence.

## The artist in a computer-mediated culture

> Research agendas can determine the flow of the future. For example, researchers working on ideas such as ubiquitous computing (making objects intelligent and aware of their surroundings) are working on more than just new products. They are transforming the primordial relationship of humans to inanimate objects. The way that research unfolds will become a crucial part of our cultural heritage.[12]

11 Sullivan, *Art Practice as Research*, p. 23.

12 S. Wilson, 'Research as a Cultural Activity', <http://userwww.sfsu.edu/~swilson/>. (A modified version entitled 'Why I Believe Science and Technology Should Have No Borders' was published as an opinion editorial in the *Times Education Supplement* on 7 December 2001.)

Wilson's argument is that it is not enough for artists simply to use new digital tools; they must be at the forefront of pioneering research in the development of technology. An artistic enquiry can have both practical and philosophical consequences and therefore can have significant impact on the individual in society. The Social Sciences and Humanities Research Council of Canada defines an artist-researcher as a member of the faculty of a Canadian post-secondary institution whose work involves research and the creation of works of art. In their terms, research/creation refers to any research activity or approach to research that forms an essential part of a creative process or artistic discipline and that directly fosters the creation of literary/artistic works. As with all research councils, applications have to show that the research has clear research questions, offers theoretical contextualization within the relevant field or fields of literary/artistic inquiry, and presents a well-considered methodological approach. Peer standards of excellence have to be reached and outputs must be suitable for publication, public performance or viewing. I cite this here as my engagements with two research projects with Hexagram (Institute of Media, Arts and Technologies, Montreal, Concordia University, Canada): Narrative: Textiles Transmission and Translations (2004–2008) and Wearable Absence (2006–2008) were driven by these criteria. Textile Transmissions and Translations is a research project that takes advantage of the ability of fabric to impart meaning through material and electronic languages, by combining a creative approach to the textile arts with technical innovations in circuitry and wireless transmissions: exploring ubiquitous computing, mobility and interactivity through the introduction of electronic devices into fabric structures; creating animated displays on the surface of cloth, in order to extend its dynamic, narrative abilities; and developing a transitional space in which meanings are altered and textiles are invigorated into new patterns of discovery. The research focus of Wearable Absence centres on offering a unique vision of future textile technologies situated in a personal, social and cultural context. Wearable Absence presents a mini-frame in which clothing becomes the catalyst and filter within the process of retrieving rich media content according to biological data. The collaborative team is across two universities, Concordia and Goldsmiths, and involves engineers, computer scientists, visual artists, textile designers, writers and five PhD students across the same disciplines.

Objects and artefacts that are at a stage of early prototype between Hexagram and Goldsmiths can produce new types of knowledge and understanding and are the significant outcomes of practice-based work. In our case, this research is at the forefront of textiles and technology. From another perspective, new knowledge-based research findings and artistic works can also be created in parallel. I have an emerging practice with Dr Tim Blackwell, my colleague in the Department of Computing at Goldsmiths: *Swarm Tech-tiles* or A *Sound You Can Touch* is an ongoing interdisciplinary collaboration, the aims of which are an exploration of both visual and sonic texture in the abstract, as well as the development of new textile designs and sonic works. Textile images are mapped and taken from scans of complex weaving patterns generated by the jacquard loom into aural structures that can be played to an audience, offering a totally new point of entry into the world of virtual textiles. The resulting work incorporates cutting-edge digital technology to transform the virtual warp and weft of mutating textile patterns into sonic improvisation, creating an environmental installation where the immediate haptic and abstract aural qualities of the material are made available for a multi-sensorial experience. Our method is based on a complex Swarm abstraction animation process that unpacks the fascinating micro-texture of virtual surfaces, investigating the space in between accessible dimensions of material objects where alternate worlds of experience might reside. The work probes notions of creativity and glimpses into possible meanings and connections between mobile bodies of sound and image texture in time and space. Our collaborative process continues to push the boundaries of cross-disciplinary artistic practice, embracing new technologies combined with live performance.

One of the issues for individual visual arts practice has been in the definition of contexts and discourses not widely known or understood by the general public, let alone the 'old' university sector. I have found that working in collaboration across disciplines and in multidisciplinary teams has helped enormously in this process, as evidenced by the work I have pursued with Tim Blackwell. The attention to disseminating and communicating the potential understandings of the artefact may assist artistic practice more generally and not only that pursued under the auspices of research: in our case the work we generated is articulated through journal papers and in live performance. Live performance and the virtual textile patterns that are generated are produced in real time and do not result in an object or artefact.[13] This concept is taken further by Nigel Krauth, who, in his article 'The Preface as Exegesis', states: 'the position of the creative researcher in the culture of the twenty-first century is not oracular: it is interactive'.[14]

This is not to say that the object or artefact is unimportant. I still have an individual practice as well as a cross-disciplinary one. My point is that an additional explanatory space is provided as well. Practice-led research can be seen as an exercise in 'consciousness raising'. This can ensure that the artist as researcher, within a computer-mediated culture, is one who operates as a creative practitioner and is made visible within the surrounding culture. Another voice, that of 'alternative' logic of practice, is accessible and heard. Environments and contexts for production have shifted to become more complex, discipline boundaries have become increasingly blurred. Christiane Paul describes these environments.

13 T. Blackwell and J. Jefferies, 'Swarm Tech-tiles', in E. Rothlauf et al. (eds), *Applications of Evolutionary Computing* (Evoworkshops, 2005), pp. 468–77 and T. Blackwell and J. Jefferies, 'Collaboration: A Personal Report', *International Journal of Co-creation in Design and the Arts*, 2/4 (2006): 259–63.

14 N. Krauth, 'The Preface as Exegesis', *TEXT*, 6/1 (2002).

> The creation process of digital art itself frequently relies on complex collaborations between an artist(s) and a team of programmers, engineers, scientists and designers … Digital art has brought about work that collapses boundaries between disciplines – art, science, technology and design – and that originates in various fields, including research and development labs and academia.[15]

The area of visual arts practice where this creation and complexity is most prolific is, as I briefly sketched through some of my own recent endeavours, at the intersection of art, science and technology. Artists are not only exploring the digital world but, like Stanza and Mick Grierson, have become rather brilliant programmers, generating new creative tools and software positioned at the forefront of new knowledge across disciplines. This knowledge is not based on the production of what was once a traditional object or an artefact of studio-based production, but rather this knowledge is produced in a conceptual space, infused with cultural discourse, open and flowing with negotiated meaning. The Internet becomes rather like a definition of installation. It comes alive, so to speak, as Valovic points out in Sullivan's lucid book, when the viewer interacts with the work.[16]

The shaping of digital and human systems by mutual interaction is one of the driving features of artists who have emerged from a visual arts background. They are now working in computer-mediated studios, development labs and academia. Margot Lovejoy comments:

> A flexible, nonlinear, interactive system or structure, one designed and coded within linking capabilities which allow the viewer to make choices in moving along different paths through the work. With interactivity, readers, viewers, listeners can pass through the boundaries of the work to enter it. This puts them in a position to gain direct access to an aspect of authoring and shaping the final outcome of a work in a way that has never existed before the advent of the computer.[17]

I am not suggesting that artists give up total control to the viewer, although interactive tools have been devised which can be circulated and used by any number of creative individuals, but rather that the artist-researcher working within the digital opens an array of opportunities for us to explore multi-modal (image, sound, text) combinations as sources of new knowledge and understanding.[18]

---

15 C. Paul, *Digital Art* (London: Thames and Hudson, 2003), p. 22.

16 Sullivan, *Art Practice as Research*, p. 155.

17 M. Lovejoy, *Postmodern Currents: Art and Artists in the Age of Electronic Media*, 2nd edn (New Jersey: Prentice Hall, 1997), p. 165.

18 S. Wilson, *Information Arts: Intersections of Art, Science, and Technology* (Cambridge, MA: MIT, 2002). Having reviewed and provided detailed accounts of 250 international artists, Wilson concludes that research is a cultural activity in which outcomes

## The artist in a computer-mediated culture: Inside the academy

In 2005 the Arts and Humanities Research Council initiated a review of practice-led research in art, design and architecture which included the Fellowships in the Creative and Performing Arts, first introduced in 1999. The purpose of the review was to develop a 'comprehensive map of recent and current research activity in the area' and the context in which to place it.[19] The AHRC, in the guidance on Fellowships in the Creative and Performing Arts, requires that some form of critical written analysis must accompany creative practice. Rust et al., the authors of the review, went so far as to suggest that this requirement implies that a reflective component to the research must be included. However, 'Taken in isolation this may suggest that some creative work might be "converted" to research by a suitable accompanying text.'[20]

They conclude that although the art community has received a substantial proportion of funding, the Fellowships do little to advance the academic community in the subject. They go so far as to suggest that the scheme neither 'promotes advanced researchers nor facilitates advanced practitioners to develop into leading researchers'.[21] There is no comment on the questions of the impact of technologies on research processes and art practices. If we can now accept the digital and interactive scenarios of art making then we engage with an exciting hybrid combination of databases, algorithms, virtual textiles, reactive feedback systems and software/coding, multisite telematic performances, net art and sense cityscapes. I briefly described the impact of these scenarios on my own practice earlier in this chapter but what I want to pursue in the last part of this chapter is how the research endeavour has provoked new practice-based approaches.

> Practice-based approaches can only aid traditional academic research, whether or not traditional research is eroded or renewed. Computer-mediated culture has had an enormous impact on practice research in that it makes both the generation and dissemination of outputs far more simple and cost effective. The proliferation of data helps to make all sorts of research more coherent and successful by providing a greater level of resources, including social networks and the exploration of audience/user studies. Computer-mediated culture puts us in touch with each other, so that research and evaluation can be fast, efficient and open.

---

are seen in terms of human exchange and as such are not the province of particular domains or privileged methods of inquiry. This view is upheld by Sullivan, *Art Practice Research*, in his appraisal of the responses of artists within a computer-mediated environment (pp. 156–7).

19 C. Rust, J. Mottram and J. Till, *AHRC Research Review: Practice-Led Research in Art, Design and Architecture* (February 2008), <http://www.ahrc.ac.uk/apply/research/sfi/ahrcsi/research_reviews>.

20 Ibid., p. 65.

21 Ibid., p. 58.

As in the above extract, Mick Grierson is very clear about the value of the AHRC Fellowships.[22] Mick Grierson is AHRC Fellow in the Creative and Performing Arts at Goldsmiths and an experimental artist, musician, filmmaker and researcher specializing in the field of real-time audiovisual composition, installation and performance. Drawing on the rich history of experimental film and sonic art, he fuses technical and artistic approaches to produce a combined contemporary audiovisual experience. He works with his own software, employing techniques from artificial intelligence, chaos theory, neuroscience and psychology to explore reactive audiovisual feedback systems, creating experimental musical film experiences. These experiments continually evolve and respond to the performer, reflecting and affecting the environment. He is currently working on cognitive and structural approaches to contemporary computer-aided audiovisual composition as part of a three-year AHRC-funded research project at the Electronic Music Studio, Department of Music, Goldsmiths. In an email conversation with me, Grierson expressed the view that the AHRC funding scheme came at a time in the development of contemporary arts practice at which opportunities for new research within the arts and humanities should be more widespread.

One of the long-term objectives that we all share is to build research networks and projects that are multidisciplinary, allowing artists as researchers and technologists alongside computer and social scientists to explore what new concepts of value and discourse in creative practice can be generated.

## Stanza: The Emergent City

Stanza is the AHRC Creative and Performing Arts Fellow, Goldsmiths Digital Studios (GDS) (2006–2009). He is a London-based artist, who specializes in net art, multimedia and electronic sounds. The research he is undertaking as part of his AHRC Fellowship with GDS is around the Emergent City, an exploration of data within cities and how these can be represented, visualized and interpreted.[23] The underlying theme or main research question is concerned with the organic emergence of the 'city'. Cities clearly have a character, but do they have a 'soul'?

22 Email conversation between Mick Grierson and Janis Jefferies, 17 April 2008. As Simon Biggs points out (2006), music has had a long tradition of practice-based research in UK academia, as most music departments can be found in old universities, though they too have benefited from the formation of the AHRC (1997) as there was no other research council formally to support their research. In July 2008 Grierson was also appointed as 0.5 lecturer in Computing to co-direct the Creative Computing BSc/MSc programme.

23 Goldsmiths Digital Studios is dedicated to collaborations among practising artists, cultural and media theorists, and innovators in computational media, who are expanding the boundaries of artistic practice, forging the future of digital technologies and developing new understanding of the interactions between technology and society.

If so, then what can be determined about the validity of the city experience in relation to its 'character', or 'soul', or 'spirit'?

Data from security tracking, traffic data, and sensor data for environmental monitoring can all be interpreted as a medium to make process-led artwork.[24] The research project also involves collaboration and mentoring from Professor Michael Keith, Director of the Centre for Urban and Community Research and head of Sociology at Goldsmiths. GDS was able to support Stanza as he was exploring and expanding the uses of new technologies, making new kinds of art, effecting changes in products at their inception, providing a showcase for technology and for art, and generating new ideas whilst the Centre for Urban Studies added its considerable expertise and evaluation of ethical and social political issues surrounding data tracking and surveillance.

Stanza's research explores another dimension of how digital art and its distribution can keep pace with rapidly changing technology. The artworks he produces relate to current data flows in the environments which he monitors. [Plates 3.1 and 3.2, and Figure 3.1] The research utilizes systems that can be used to gather information via sensors in the field. This data is harnessed to visualize the urban environments and spatial representations as a dynamic real-time experience rather than as a recorded 'photograph' of the city space. He is attempting to move towards a point where the landscape is a hybridized audiovisual representation of the space, i.e. an audiovisual experience based on the sounds and sights of the city's pollution, noise and traffic data that are captured via these sensors' networks. As his artistic mentor, I have had many conversations with Stanza as to how we understand and value information. It seems reasonable to suggest that visual metaphors might simplify our understanding of data in space. Adopting visual and poetic metaphors for gathered data enables a multi-point perspective. The increase of technology infrastructure in the daily existence of a city means that technology will, more than ever, be everywhere in our environment. The patterns we make, the forces we weave, are all being networked into retrievable data structures that can be sourced for information and new forms of creative practice. Stanza analyses such patterns to disclose new ways of seeing the world and new ways of understanding it. Uses of this information and data allow rich new interpretations of the way our world is built, used and designed, now and in the future. It is possible to expand our perception of the city as a dynamic network and how this information can be incorporated to make generative artworks. Stanza's website includes all his research and

24 E.A. Edmonds et al., 'The Studio as Laboratory: Combining Creative Practice and Digital Technology Research', *International Journal of Human-Computer Studies*, 63/4–5 (2005): 452–81. This paper is concerned with the nature of creativity and the design of creativity-enhancing computer systems. The notion of the studio as a laboratory in the field is introduced and a new methodology for systematic practice-based research is presented. See also L. Candy and E. Edmonds, *Exploration in Arts and Technology* (London: Springer, 2002).

**Figure 3.1** ***You are My Subjects***

*Source*: With permission from Stanza. Real time CCTV Internet art, public display, 2006.

findings. It houses all creative developments, a gallery, and actual real-time data gathered from my sensor network (for others to see or to use). When we were mapping out the original application to the AHRC, one of the most challenging tasks was indeed to address specific research questions and to redefine how art, artists and audiences might engage with the results.[25] The questions we discussed were how we could consider data (CCTV, traffic, weather, people movement) as a medium for artistic practice. How could this information be meaningfully represented as artworks? How could this data be displayed in new and original ways? Do the results create new ways of understanding the city space? [Figure 3.2] Nonetheless, Stanza believes the discourse, conclusions and visualizations arising from this research programme will be of value to a wide general audience. His rich and comprehensive website provides access to his research findings, the developing artworks, and the code and technical details. Stanza's diary is updated on a monthly basis and provides continuous progress reports relating to developments in his research. We are aware that the technical

25 The Research Office at Goldsmiths and particularly Lynda Agili offered critical insight and support throughout the application process. The value of such expertise can make or break an application.

components need to be balanced with a stronger theoretical critique of the field and the project itself, and in our mentoring sessions one of the contributions that Michael Keith and I make is to support his writing. Stanza's ongoing text, a diary in process, acts as a bridge to his many research projects and is supported by his research blog: <http://www.theemergentcity.com>. We agreed that the mix of procedural development and evaluation through practice was the preferred research methodology. For example, if the technologies change, Stanza adapts to them, thereby keeping his research path open. He does this by visiting other experts and by testing how some of the ideas from the AHRC Fellowship can be developed in a gallery context. For example, in February 2008 Stanza undertook a trial residency at Plymouth Arts Centre, producing an exhibition called 'Visitors to a Gallery', which is part of a series of process-led experiments in data visualization using sensors and CCTV.[26] What happens during the process of visiting the gallery and data space and what audiences do are key questions (and experiences) that Stanza addresses. At the end of the Fellowship, a final public artwork will be presented and made available for touring.[27] This process requires collaboration and curatorial partnerships with media-based organizations such as Watershed, Bristol and SCAN.

At the end of his second year of his Fellowship, Stanza had investigated further the computer science and engineering aspects of wireless networks. These have been added to his core creative output, integrating other systems and physical objects. What is important is that the work can be live and not restricted to dissemination via recorded archive; some significant access problems are being erased by open source approaches and knowledge sharing networks.

When I asked Stanza, Mick Grierson and Tim Hopkins what impact the AHRC fellowships had made on their practice and performance-based research and whether or not the scheme had enabled them to pursue their careers as active academic researchers they were positive about the opportunities, networks and collaborations offered. Grierson was especially vocal in concluding that 'Within the sciences, fellows are given the right to apply for further funding as Principal Investigators beyond the term of their fellowship but this is not the case with those offered by the AHRC.'

---

26 <http://www.stanza.co.uk/gallery/index/html> and 'Visitors to a Gallery', <http://www.stanza.co.uk/cctv_web/index.html>.

27 Helen Sloan, Director of SCAN, and I were in conversation throughout April and May 2008, exchanging views on the issues of interdisciplinary research across arts and science and on how curating can be a form of practice-based research in respect of working with artists and new technologies, defining exhibition themes and events across different networks. These ideas were articulated by Helen Sloan at the *taxi-to-praxi* workshop organized by Armin Medosch and Adnan Hadzi (PhD students at Goldsmiths in Arts and Computational Technology and Media and Communications respectively), Goldsmiths Digital Studios, 21 April 2008. The debate can be accessed via <http://www.thenextlayer.org>. Helen was responsible for collaborating with Stanza for his exhibit *Inner City and Biocity* as part of the Venice Biennale 2007, New Forest Pavilion.

**Figure 3.2** ***Globals***

*Source*: With permission from Stanza. Live data globe, net art, coded visualization, 2004.

Their research is well regarded and has a direct impact on the wider artistic and research community and on contemporary culture more generally, for example on commercial frameworks such as industry, including galleries and performance venues. The challenge is how to sustain innovative practice within the academic research environment, respecting its differences but recognizing that higher education institutions have gained a great deal in terms of research. The institutions have to acknowledge that in the twenty-first century ideas and creative practice are not just limited to visual arts but have rapidly expanded though a computer-mediated culture. Understanding who we are, what we do and what we know is changing faster than we can keep up. Artists engaged in the academic world, where their practice has primarily developed outside it, effect and are affected by encounters with other disciplines in unexpected ways, and this is precisely where the richness of research culture resides. What has emerged is a new type of multidisciplinary worker or collective: the artist as researcher, participating at the same time in artistic practice and the development of technologies that are shaping our society. The writing, often anecdotal and autobiographical, that contextualizes this work can be viewed as mapping the intricate relationships generated by practitioner-based research, questioning in what sense it is the best way to understand our relationship with traditional research fields.

**Acknowledgements**

With enormous thanks to Stanza, Mick Grierson, Tim Hopkins and Helen Sloan.

# Chapter 4
# A Conversation about Models and Prototypes

Jane Prophet and Nina Wakeford

## Introduction by Nina Wakeford

Rapid prototyping (RP) enables the creation of physical objects from computer-aided design packages. In its cheaper versions, it is sometimes known as 3D printing, and the RepRap, cited below, even promises a machine which can replicate a full version of itself. In her recent projects, Jane Prophet has pioneered the use of RP in art practice, and in this conversation she explains how she began to engage with the technologies of simulation and virtuality, and how this led her to investigate the utility of RP for making art. In our edited conversation below – which itself began as a prototype-like interview – we were concerned to explore the ways in which RP is connected to other forms of computer and digital culture, as well as to contextualize the work in terms of other perceptions of models and modelling. Many readers will be familiar with Jane's *TechnoSphere* project, and she begins by reflecting on the politics of such exercises in simulation. The conversation also highlights the very different ways of understanding the potential of the virtual, particularly when an artist becomes involved in interdisciplinary collaborations with mathematicians and scientists. In contrast to Jane's earlier Internet-based work, the use of RP brings her back to the importance of the material properties of her objects, such as the effects of silver plating. RP appears to be a form of maquette making, or a way of testing something that might have a much larger final form. Yet Jane notes that it is only after the RP object is shown as an artwork that she might begin to think of it as a maquette. Her close collaboration with teams that are trialling RP technologies allows her to imagine new forms of RP machine which might better serve the artistic imaginary, although she acknowledges that RP is much more likely to be found in a design department.

Nina: When did you start using models?

Jane: I was at Sheffield Polytechnic in the eighties in Communication Arts, where I used video, film and sound. Through working with my tutor, Fran Hegarty,[1] a performance artist who uses video, I became interested in, and attentive to, the body in space. She trained me to think that every single thing in a room, and what happened across time, was very, very important. In light of that, I would describe my practice as being one that looks at structure and space. Sometimes that is quite literal, for example exploring the physical structures of a building, but equally it might be about anatomical structures such as the heart, or social structures. There can be an interesting overlap, or a kind of slippage, between how we think of a structure in a physical way, and the fact that the structure is iconic. Culturally, its meaning moves around. I think Fran's legacy can be traced through my practice, as a lot of my works perform in some way – they have a time-based element. I am not *in* them, but the concerns of much live art are still in the pieces. I didn't make models at art school, because I was making performances or videos. Instead I sketched, very basic 'stick people' drawings, to plan pieces, like storyboarding. I also sketched in a less representational way. I once made a piece about the docks in Hull when they were in decline. As first year students we were expected to draw boats, but I hung out in the dockers' café with unemployed dockers who still went to the café every day because that was their routine, their structure. I got to know them a little and a couple of them took me around huge, derelict, cut-up trawlers. During those walks I accumulated samples in plastic bags (sacks of flakes of rust, dried pigeon shit, dozens of discarded welders' gloves) and I'd lay them out to make compositions, but it wasn't representational sketching. I think now, when I work with models, that there is a similarity to that gathering of impressions. I am not necessarily trying to represent something literally, I am trying to capture the essence of a structure.

Nina: What were your first experiences with what we might think of now as digital art? How did that relate to your thinking about structure?

Jane: I did my MA in Electronic Graphics at Coventry University, which was strange as I was really a technophobe. It was incredibly frustrating, I didn't produce anything in that 18 months that I liked. It was all awful looking and clunky. But I did learn something about computers. The thing that kept me working with that technology was nothing to do with the images I could make, it was about structure. Ideas for performances and installations were described in circles that overlapped – like Venn diagrams – ideas described as a collection of single words, each set in

1 Fran Hegarty is Emeritus Professor of Sheffield Hallam University. Her work as an artist spans three decades, at times concerned with received ideas of cultural and national identity, with emigration, with the female body and mortality. She works with video, audio, photographs, drawing and installation, exhibiting worldwide. <http://www.brighter.org/franceshegarty/index.html>. (All URLs current at the time of writing.)

a circle that connected to other circles with lines. When I started learning about computers in the late 1980s, the thing that had the biggest impact on me was hypertext. The idea of non-linear connections, and jumping between things, was exciting. Recognizing the power of associative connections was hugely motivating to me, because it was what my sketchbook diagrams represented. It was the way that I could structure information that made me stick with the technology.

Nina: During your MA, did you produce anything that you would now think of as making a model?

Jane: I made a lot of concept maps, and my thesis was a model for thinking about non-linear structures. I had to submit a bound thesis, so I placed little tags round the edges so you could flip through the thesis in a more non-linear way. I did learn 3D computer modelling, which at that point meant typing in x, y and z coordinates. I had a matchbox on my desk because I was trying to understand 'z space', which you would have thought that I would understand, because I did performance and worked in big spaces. But z space was very challenging to me. It seems strange now, but screens really meant two dimensions to me then, and working with a virtual third dimension was difficult. Nothing I made was of any interest, but conceptualizing a z space: a three-dimensional virtual space, was pretty radical. At Coventry I met Gordon Selley,[2] who I subsequently worked with for many years. He worked in the Computer Science department, modelling natural phenomena – fog, light, foliage – for flight simulation, for his PhD, and I thought it was the most bizarre thing to do. I understood why flight simulators need to have landscapes and weather effects that look real, but as a core engagement I thought it was strange. The whole computer graphics industry at that time (the late eighties) seemed to be driven towards photo-realism. In the arts at that time photo-realism was a movement that you laughed at. Photo-realistic painting seemed really nostalgic. On my foundation course I was taught by the Super Humanist painter, Neil Moore. Whilst I respected his paintings, what I appreciated about them was not their photo-realism, but their symbolism.

Nina: So the idea that simulation was the driving force of much early computer graphics was less interesting than how it opened up the conception of a different space?

Jane: It was about space and time. I found it engaging and yet struggled to understand what that drive to photo-realism in computer graphics was about. Lots of people at Coventry University were interested in photo-realistic rendering and photo-realism in computer graphics and I used to question this. I concluded

2 Gordon Selley is a computer graphics programmer who collaborates with Jane Prophet on a number of projects including *TechnoSphere*, <http://en.wikipedia.org/wiki/TechnoSphere>.

that there was a nostalgia for the sorts of landscapes they were modelling – they were not modelling local parks, but iconic sublime landscapes, such as the Grand Canyon. These simulations, from an art historical perspective, were a contradiction in terms. It was important to me to think about *why* they were trying to reproduce these incredible sublime landscapes, which of course are no longer sublime to us.

Nina: What strikes me is the politics of these representations, and it reminds me of the early 'virtual world' industry in San Francisco in which developers were producing architectural visualizations that resembled the urban infrastructure of Silicon Valley, perhaps because that was what they saw from their windows. Perhaps we might say that these models reflect a certain kind of politics? Do you think that at Coventry University they were ever aware of that kind of politics?

Jane: No. Although I think that there is now a willingness to engage in conversation, and an open-mindedness that wasn't there then. It is still a predominantly male industry, and the language associated with photo-realistic real-time graphics – the words used but also the tone of voice – focuses on the thrill of flying over these landscapes and seems very gendered. Obviously there are thrills in 'shoot-'em-up' games, but this is a different thrill – the exhilaration of flight, of flying over landscapes or sweeping down and seeing trees and rock formations and suchlike. I think the language is similar to accounts by early aviators. It alludes to a sense of conquering nature, surveying it from a great height and at speed. In the simulations the landscape is symbolically 'mastered' because it has been deliberately designed, but it is also conquered through the experience of simulation, in a way that you cannot now conquer the real landscape. Now, when you fly over the Grand Canyon and see a glass walkway sticking out of it, you can't fantasize that you are the first. If you go to Mount Everest there are crowds, discarded oxygen cylinders and litter. The opportunity to fantasize that you are conquering, that you are the first, is eroded. It is as if you have to simulate the experience of claiming this kind of so-called natural territory for the first time, by reproducing it on a computer.

Nina: There is a history of the model being central to projects concerning enlightenment and the mastery of nature. For example, Simon Schaffer has written about attempts in the 1770s to create a model which showed that a stingray mimicked an electrical instrument.[3] When the experimenters demonstrated that they had control of the model, this confirmed their philosophical claims about the way in which nature behaved. So, showing control over nature via models seems to be a longstanding practice. How do you think about nature within the politics of simulation, of representation? [Plates 4.1 and 4.2]

3 S. Schaffer, 'Fish and Ships: Models in the Age of Reason', in S. de Chadarevian and N. Hopgood (eds), *Models: The Third Dimension of Science* (Stanford: Stanford University Press, 2004), pp. 71–105.

Jane: I think about nature in a very particular way. The sublime experience doesn't exist unless there is a figure in the landscape to provide the scale: the subjective experience of the sublime is embodied. You have to be there, you have to be prompted to think of the infinite 'god' that produced this thing, to feel terrified, and then you have to feel comforted. To me that's an essential part of the sublime experience, to have fear, and then reassurance (which is what happens in games). We know we will crash and burn, but then we press restart and it is all okay. So, I think that figure in the landscape is really important. I am also interested in the model as something represented at a different scale, or in a different form, to what is out there in the world. Equally, I am interested in the concept of the model as an ideal. Most simulated computer environments are idealized.

Nina: Do you think of models as forms that have the capacity to condense?

Jane: Yes. As a layperson who has been party to teams discussing the design of a flight simulation war game such as *Back to Baghdad*, I've noticed that the use of satellite data is considered important, because it has veracity and it is photo-realistic. It is equally important to get to the essence of something. It is also about erasure, leaving out what might be termed superfluous detail. Detail *is* kept in, but it is detail that accentuates or amplifies a particular position, experience or viewpoint. Similarly, photo-realistic paintings could be of highly reflective surfaces such as, for example, a composition of mirrors and glass objects, but details such as finger marks were not usually included. A computer simulation will not generally attempt to simulate details such as rubbish or blowing leaves, partly because dirt is computationally expensive, and difficult to model. Anything that is included in a computer simulation must, from a design point of view, add to the 'myth' and build a particular position or aura.

Nina: So, models are necessarily idealized in that kind of digital scenario?

Jane: I think so, yes. They are also unrealistic in all sorts of ways. I am fascinated by the complexity of the algorithms needed to define the appearance of things. In 3D computer space there is no atmosphere, dirt or gravity. All have to be simulated. There is no reason why one object in a virtual space should not pass through another, and to put in collision detection to make sure an object doesn't do so is expensive computationally. Modelling the physics and dynamics of the real world is onerous.

Nina: How does this relate to the way you have used models?

Jane: After my MA, the first works that I made were online. What I was interested in was typified in the artificial life project, *TechnoSphere*, made with Gordon Selley in the mid-nineties. I looked at the structure, or a conceptual model, of the World Wide Web. What was it, what did it do, how did it function, and how could I use it to make an artwork? Back then the Web was really not good for presenting big two-dimensional images. Owing to connection speeds and technological limitations, it was even more difficult to view moving images or three-dimensional rotatable objects. However, the Web was good at presenting blocks of text. The challenge to me as a visual artist was to use this quality to make something that gave a sense of time passing, and that had a recognizable cause and effect. This led me to work with Gordon on *TechnoSphere*. In gaming, at that time, there was a tendency for a small action to lead to something dramatic, like an explosion, or carnage on a great scale. What was interesting about *TechnoSphere*, which we gained from user feedback, was that people were engaged with the project although there was very little immediate drama. Users chose two-dimensional images to build creatures (heads, bodies, legs, wheels), gave them a name, put them into a virtual environment that you couldn't see, and then got email messages back from them. There was so little reward, but I realized that what was important was the connection these people felt to the virtual creatures. The reason they felt connected was simple, and was the thing the net did really well – communication via email.

Nina: Would you describe *TechnoSphere* as a model?

Jane: Yes. It was a partial model of a carbon-based ecosystem, and we were upfront about its limits. Even so, it sometimes led to unexpected things. We received some enraged emails from creationists whose children had made a creature at school, using their parents' email address as the line of communication from the creature. When these people saw that we used words like 'evolve' on the website, they were outraged. I hadn't heard about creationism in 1995, but I learnt about it very quickly! We didn't make *TechnoSphere* to offend people but the idea that it could trigger off a heated debate was fascinating. *TechnoSphere*, in a small way, was a model for all sorts of people about what you might be able to do online. Because we did genuinely try to build users' suggestions into iterations of the project as it evolved, people felt a sense of ownership and a sense of openness.

Nina: Yet people often expect models to have some kind of stability …

Jane: That is right, but I think that is a pre-Internet idea – that once you make something, it is there in perpetuity. By contrast I see much of my work as ephemeral, because I work online and with digital technologies that are inherently unstable. I am comfortable with that, perhaps because of my earlier experience with performance art, where a performance could often be a one-off event. All that is left is documentation, or what people remember about it. *TechnoSphere* is a good example. It was less distressing for me to take *TechnoSphere* offline than it was for some users. I think ideas about models are that they will *always* be there. The idea that you can model something and then lose the model is unnerving.

Nina: The way you talk about your work calls to mind Celia Lury's description of new media objects as 'performative'.[4] Her object is a brand, but her formulation of a new media object as a 'dynamic platform or support for practice' seems relevant to this discussion. Taking this further, perhaps you are talking about the model being an event, so it is not just a performative object but has the qualities of an event in it. Does this proposal of model as event resonate with your later work, after *TechnoSphere*?

Jane: In 2002 I began working with Neil Theise,[5] a stem cell researcher and liver pathologist, on a project called *Cell*. He taught me about stem cells and how they behave, and I talked to him about art, complex systems and media. He had never heard of complex systems and he wasn't convinced about them until I talked to him about ant walk theory, which engaged him because as a kid he was fascinated with watching ants. Working with the mathematician, Mark d'Inverno,[6] we modelled stem cell behaviour. It took four years because we each thought so differently about modelling. The language that we used to describe actions, events and objects also differed. Mark d'Inverno and I wrote, in plain English, a model for the behaviour of stem cells in the adult human body. The supposed 'hard' medical science was much softer – more open and porous – with more 'unknowns' than we had expected. This was a challenge, and drew attention to our differences. As an artist, I found the lack of answers surprising, but I could cope with it. However, Mark found this difficult. As a mathematician, he sought answers – answers are needed in order to develop a computer model of how something behaves. The model is a set of rules that define behaviour, which is what we put into practice in *TechnoSphere* in a very simplistic and imprecise way. The *Cell* project had a more 'hard science' approach and I was shocked that medical researchers were proposing, and proving, hypotheses about the way stem cells behave in a complex living organism, usually a mouse, by killing the organism and slicing it into thin sections – to me it was a crazy way of determining how a complex system works. To Neil, it was *the* way to do it. Our proposal seemed completely crazy to him. We said 'Why don't we try and write the rules of behaviour as you think they are, with you acting as a kind of collective stem cell research community, and then

4 C. Lury, *Brands: The Logos of the Global Economy* (London: Routledge, 2004), p. 6.

5 Neil Theise MD is a diagnostic liver pathologist and adult stem cell researcher in New York City, where he is Professor of Pathology and of Medicine at the Beth Israel Medical Center at the Albert Einstein College of Medicine. His research revised the understanding of human liver microanatomy which, in turn, led directly to the identification of possible liver stem cell niches and the marrow-to-liver regeneration pathway. He is considered to be a pioneer of multi-organ adult stem cell plasticity and has published on that topic in *Science*, *Nature*, and *Cell*, <http://www.neiltheise.com/>.

6 Mark d'Inverno is Head of the Department of Computing at Goldsmiths, University of London. His main area of research is the theoretical development of intelligent agent systems and their practical application, especially in an interdisciplinary context, <http://www.doc.gold.ac.uk/~mas02md/>.

run them on the computer and watch these things in real time and 3D and see if what is happening in the simulation bears any resemblance to the sliced-up two dimensional bits of a mouse.'

Nina: So at that point you were suggesting what would have been a fundamental shift in the discipline?

Jane: Yes, it was. We assumed that people in this area of medical research would model using computer simulation. It seemed obvious to us: not as a replacement for animal research, and not from an ethical point of view, but because there are ways in which you can monitor a simulation that are not possible in a living organism, because there is currently no way to actually track cells in a living body without killing it, which is a fundamental problem. I am still surprised that there is not more simulation and modelling taking place. I think it is something to do with the cultural differences not just between established research techniques, but in what it means to test and model.

Nina: What kind of challenges does this present for artists?

Jane: With some large or complex sculptures there is a degree of engineering-level modelling that happens on the computer – for example, in stress testing. But beyond this, computer models are predominantly concerned with visualizing the finished work, whereas in biology and medicine the appearance of things is much less important. Behaviour is the focus. So, we are starting from different places. Maybe modelling in architecture is half-way between the two.

Nina: Albena Yaneva has highlighted the gestural knowledge of architectural models; maybe that is a point of convergence.[7] She has also pointed out that contemporary architecture needs these non-verbal and non-textural ways of communicating. So architecture is perhaps a good place to begin to think about models. However, there is also the sculptor's maquette as a preliminary concept of something larger-scale. Are there other places to look for ways of using or thinking about models?

Jane: Much of the activity of an artist is what we call 'fiddling about'. Many of the results of that activity get thrown away. For me that 'fiddling' is often about developing a model as a tool for thinking, so it might be concept diagrams, or I might think of a shape and cut it or tear it out of paper, and fold it as I think about it. Most of those things that get thrown away are invisible, it is so intuitive it is almost invisible work. I think that many artists and designers are thinking *by* making something with their hands. Whether or not that has any trace in the final outcome

7 A. Yaneva, 'Scaling Up and Down: Extraction Trials in Architectural Design', *Social Studies of Science*, 35/6 (2005): 867–94.

is another story. Engineers don't work in quite the same way. They tend to have a design, test it on the computer and then implement it and build it on a machine. But I work with one engineer, Alain Antoinette, a manufacturer 'one-man-band' who is multi-talented. His engineering and model-making skills are very much about hand-eye coordination, and he works very much like an artist. Over the last five years he has attracted more and more artist clients. I think one of the reasons that there is a good symbiosis is explained by describing a visit to his workshop recently. I went with a set of engineering drawings. I had made something out of drinking straws that had led to the engineering drawings, and what we did for two days was make things out of bits of tube. These were a kind of abomination and approximation of the engineering drawings, because we didn't believe that all the joints on the engineering drawings would work, and some didn't. Instead of talking to the engineer, we wanted to build the structure from tubes, to test our ideas and worries about the 2D model.

Nina: What did you think of the objects that you created? How did you refer to them?

Jane: We called them tests, or models. Alain said, 'We have got to build it.' So we built a small one. In the end this piece was 6m high, but we began by building things about 20cm high out of bits of Perspex tubing, to check angles and see if things would move, because this piece had to move. We realized that we were applying a lot of force to make the tubes move. That was a sketch, a rough version of our final object. Then we decided to make a bigger one from drainpipes, and that felt more like a model because that was much closer in weight, size and force to the final piece. [Figures 4.1, 4.2 and 4.3] I think model making is about deconstructing. To work out how to make my plant-like structure move, I looked for a comparison. What had I seen out in the world that folded back on itself like this? The only thing I came up with, after a long hard think, were umbrellas that fold up small. I got an umbrella, took it to pieces and we started adapting it and making its arms move differently. To make models of things that don't exist yet, we often use existing objects as models, rather than starting from scratch, taking a ready-made and cannibalizing or changing it or adapting it. I often end up with test pieces that look very 'Heath Robinson'.

Nina: So, would it be accurate to talk about this work as the creation of assemblages?

Jane: Yes, up to a point. It depends on the long-term goal. We actually had to make the complex joints drawn by the engineer, although he changed them in light of the umbrella test. He made his ideal model, the ideal joint. We couldn't make that joint work, so we radically adapted something 'off the shelf' to make it bespoke. I think that playing with off-the-shelf everyday items is an important stage in the model making process. Model making is like an event, especially

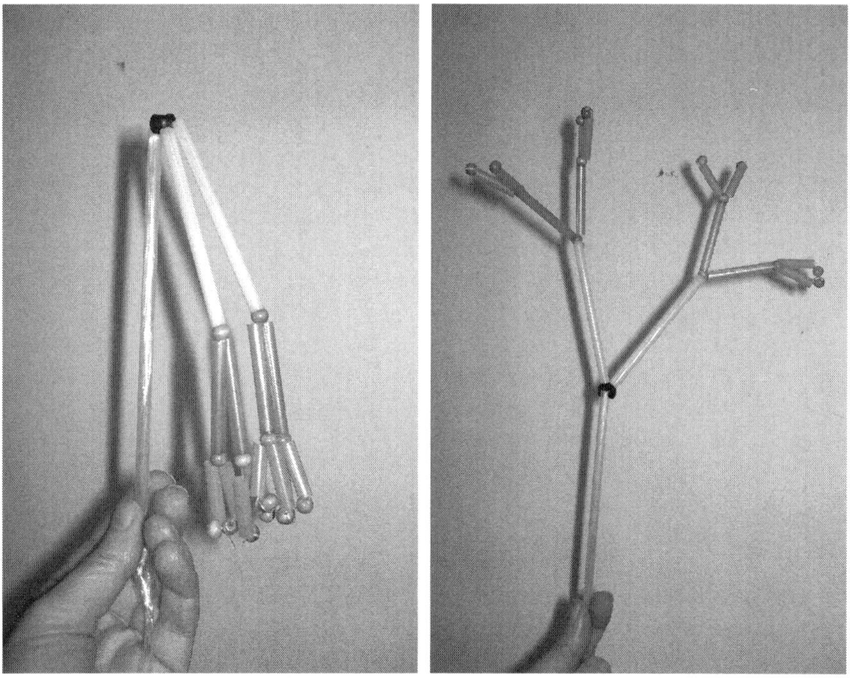

**Figure 4.1 3D model made from drinking straws**

*Source*: Jane Prophet, 2008.

if you are making something that moves. We are trying to imitate, simulate or approximate a particular movement. We find something that moves in a particular way, we study *how* this happens (like the joint on the umbrella) and then we make our own model, because the movement we want is different. There are some similarities to simulation. What is it about an oak tree that is really 'oaky'? Let's simulate that, but without modelling every detail of the branches or bark. We won't model acorns, because you are not going to see them, they are not really what an oak tree *is*. In art we see approximation as a valuable thing, but in mathematics (and I suspect in most medical research) approximation is useless, or worse than useless. It either 'is' or it 'isn't'. To approximate something in that context gives you no useful data. It might give you something else, but it doesn't help solve your particular problem, whereas approximation in model making, in art and design, often helps.

Nina: This leads us back to a conversation about model making in different disciplines, and may provide a link to your first encounters with rapid prototyping. When did you first come upon it?

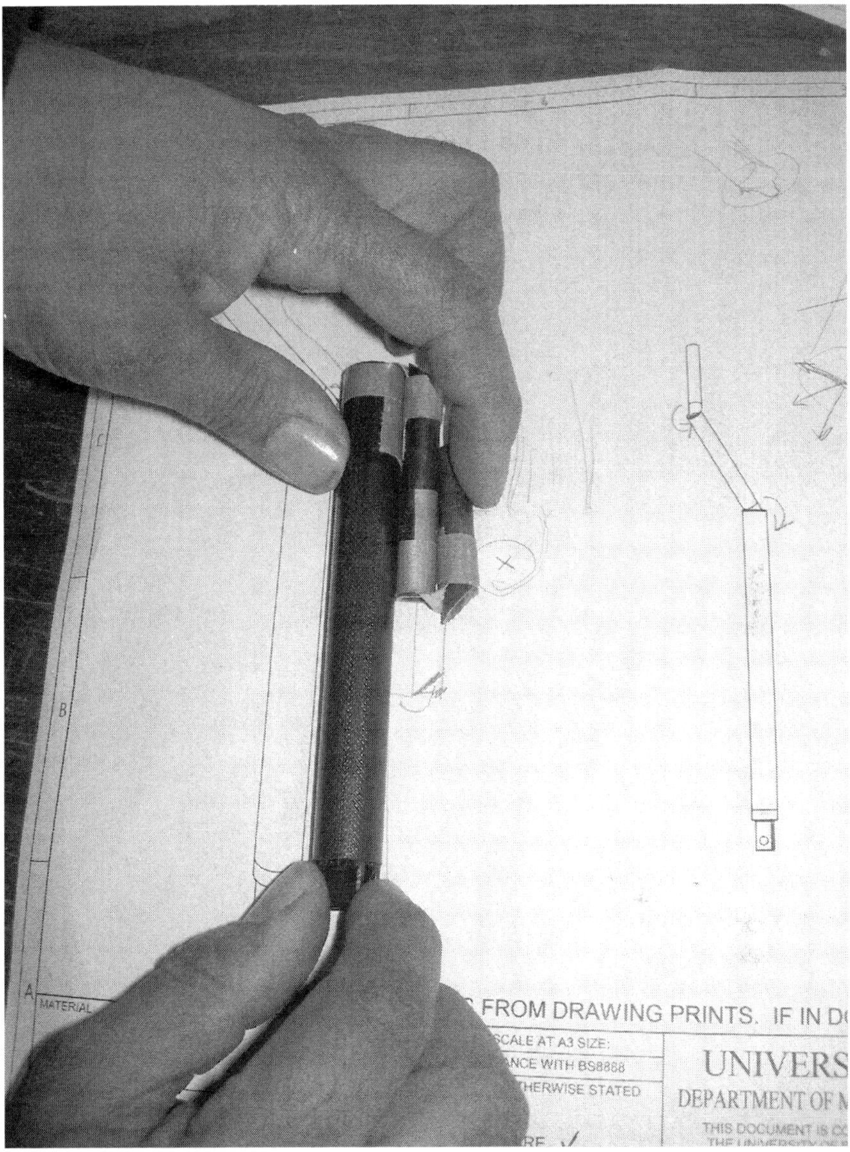


**Figure 4.2 Making model branches from CAD drawings**

*Source*: Jane Prophet, 2008.

Jane: I had read about rapid prototyping in *Scientific American* or *New Scientist*. I had seen geometric objects, things that looked like cogs, usually quite small-scale, (maybe 3 or 4cm across) generated by this means. Most of the examples were three-dimensional, but barely. I noticed that there was nothing that looked organic

**Figure 4.3 The assembled kinetic artwork *(Trans)Plant***

*Source*: Jane Prophet, 2008.

or irregular. I was interested in rapid prototyping, not to make the prototype for something, but to make the thing itself (this is now partly addressed through so-called 'rapid manufacturing'). Rapid prototyping makes real something that is virtual. This is what interested me. I wondered: 'Do you relate any differently to it when you see it and can pick it up?' I was struck by how limited the objects were, how the technology was not being, in my view, stretched very much. It was great at producing geometric shapes, so people produced geometric shapes. Well, what else can it do? Presumably it can do all sorts of things. So my focus was on 'making real' objects that had previously been virtual. The first object I made was from Magnetic Resonance Imaging (MRI) data of the human heart. I worked as artist-in-residence with Francis Wells, a cardiothoracic surgeon at Papworth Hospital Transplant Unit. Through working with him I became interested in the structure of the heart, especially how he saw a 'model' of the heart in his mind's eye, alongside what he saw in reality when he was operating. To operate you must retain multiple simultaneous understandings and three-dimensional mental images of the heart structure. One understanding is literally what you see before your eyes and feel in your hands. Another is conceptual. One of the conceptual models is of a complex, moving vascular structure. However, the complex, vascular structure shown in textbooks and via preserved specimens can never be seen in real life (unless it is plastinated and preserved as a rigid immobile object). This is because the heart is subject to gravity once it is lifted out of the chest cavity. The vascular structure, without the support of the surrounding tissues, collapses. The only time that it is possible to see the heart non-collapsed and alive is virtually, for example in in 3D MRI scans. So, using MRI data, I made a heart on a rapid prototyping machine. This was an experiment, to see how the surgical team would react to it. They had a strong reaction to it. [Plates 4.3 and 4.4]

Nina: What was the reaction?

Jane: They engaged with it. They loved it and they wanted to pick it up, which they could. At the time, rapid prototyping materials were very brittle, very fragile, and some of the blood vessels were very thin, thinner than a matchstick. However, once the heart was metal-plated you could pick it up and turn it around, and that is what they all wanted to do. You can spin virtual objects on a screen, but it is a completely different experience to actually holding an object and physically moving and turning it. The rapid prototype technology enables the production of an object with, for example, deep undercuts and grooves, which would be too complex to cast, or cut with a Computerized Numerical Control (CNC) cutter. It was interesting to see how many questions the surgical team had for each other about the surface texture when they had the object in their hands. We know a lot about the heart, but one of the things that Leonardo da Vinci proposed, although he didn't use these exact words, was that the reason the heart was such a successful pump was not because it pumped in and out like a bellows but because it twisted – a bit like wringing water from a cloth when you twist the fabric. In the last 50 years 'cardiac twist' has been proved, and that proof has come partly through imaging technology. So, there are things that we can intuit about objects that we don't know until we actually model them or image them in some way, and that is what I was interested in exploring through rapid prototyping.

After we made *TechnoSphere*, Gordon Selley and I continued our arguments about photo-realism and modelling tree structures. We wrote some plain English rules describing the way oak trees grow, and the way they look, working with Paul Underwood, head gardener at Blickling Hall, a National Trust property. We embedded these rules in algorithms, mathematical equations, and used the algorithms to produce 2D images. There is a degree of randomness in these algorithms. Each time we run them we get different structures that look treelike. We can alter the equation to apply a simulation of gravity or wind direction, and the results look like trees leaning away from the wind, on beachheads. We then applied the algorithms to three-dimensional computer space. It intrigued me to see how people would suspend their disbelief and think they were looking at 'real' trees. I made photographs, put algorithmic trees into them and printed them. People believed they were seeing a real tree, even though it was a wire-frame, so largely transparent. They said, 'Oh but it can't be real because that one is transparent. I don't understand. Did you erase part of the photograph to make the lines?' I wondered what would happen if I made 3D objects, so we took some of the computer models and eventually managed to get them in a format that the rapid prototyping machine could use to produce little tree-like objects. Some of them were impossible trees because we didn't have collision detection, so some branches blended together. It was interesting to observe people looking at the trees. They looked at them from many angles, by moving their heads around, in the same way that the surgeons moved the heart object or walked around it. The more angles they looked at the object from, the more they started to question whether it was real. Then they would have discussions amongst themselves about what 'real' meant. That, for me, was really the whole point. What is it about a structure that makes us believe it is natural and organic rather than artificial, and does it matter? For me, with model making, how does the scale of an object or a model impact on our willingness to believe it is a real organic object?

Nina: It seems to me that because the trees are less than a foot high, they reference the maquette. Doesn't this rapid prototype object reference directly back to the technology used to make it? Do these references somehow lead you in different directions?

Jane: When I made *Model Landscapes* with the rapid prototyped trees, I never thought of them as a maquette. I thought of them as a model in the sense of being 'smaller than', and in the sense of them being an idealized tree. That was my motivation. I started to think about them as a maquette *after* I had made them. They came out of the machine and they were exhibited almost immediately. When the work goes into the public world, my relationship to it changes. I am distanced from it, and one of the things about that distancing, with the small trees, was that I had a Gestalt moment. It was just like looking at a Gestalt image, when you see, for example, a wine glass, and then suddenly you also see two profiled faces. After this you can only see both images, whereas before you could have spent months only seeing one or the other. For months I had seen the rapid prototyped trees as an idealized landscape and I had been very engaged with their relationship to mathematics and rules and the modelling of tree growth. Once I saw them as objects 'out there', I suddenly saw them as a maquette, which was disturbing. I went back to the engineers at Bath University and said 'Why is rapid prototyping used to make small things?' A more important question would be 'What would happen if we scaled up rapid prototyping?' That was an important collision point for me. At Bath we had talked about approximation and rapid prototyping being about intricate detail. Every time you read about a rapid prototyping machine you find that one of its selling points is that you can get within 0.1mm accuracy. But for me, what it does in terms of making virtual 3D data real is the cool thing. I asked Adrian Bowyer at Bath University what would happen if you scaled up rapid prototyping, and he was willing to conduct a thought experiment with me. His first response was, 'But of course it probably won't be very accurate.' I said, 'No, okay, it won't be accurate, but if we made something that would rapid prototype something 40 feet high, do you not think somebody would find a use for it?' We dreamt up giant cake-icing type machines – robots that would pipe out expanding foam (or concrete) to extrude giant rapid prototyped objects, big and messy, not very accurate in relation to the computer file. For me, the making physical of something that is virtual is what is most interesting, and seeing the rapid prototype as a maquette rather than a model made me think about wanting to scale up rapid prototyping, and that triggered off a huge number of problems, questions and ideas.

Nina: Is there an emerging community or group of artists using rapid prototyping?

Jane: I think there are many more designers using rapid prototyping. I think there are all sorts of reasons for that. It uses a very uniform material, an ivory-coloured plastic (a polymer) that is brittle and not that interesting. You can now rapid prototype with inkjet coloured plaster, but plaster is very vulnerable, and the inks are not archival inks. So there are a lot of restrictions. Sculptors are used to a range of materials that have richness; different surface textures, strengths, durability. Rapid prototyping doesn't have that variety. It is also really expensive. Every time I say that, industry people say 'Oh but you can buy one for £1500.' However, unless you are going to use it a lot, £1500 is a lot of money for most artists to spend on trying something out. What this means is that there is a context, usually institutional (the places that have the equipment), within which you use the technology, and that context impacts on the take-up – on who is using it– and how it is being used.

Nina: Could you talk a bit more about the materials and the constraints?

Jane: Plastic polymer is pretty much the standard. There are desktop machines into which you can put all sorts of materials, in theory including chocolate! I think those are the machines that will enable artists to experiment and make the technology produce a wider range of objects. But you need a particular set of skills to use these things. You have to model the object in virtual space, and 3D modelling is complicated, time-consuming, and can be very boring. The irony is that the people who would probably make the most fantastic objects using the technology are traditional sculptors or people who use their hands a lot, but to generate the computer file you can't do that. I think the technologies are at an interesting breakthrough stage where they need to be able to produce things in different materials or in a material that can be coated more easily. In addition, the file formats that you can use need to be radically opened up. The technology needs to expand beyond the engineering and university workshop environment.

Nina: Can you see rapid prototyping being used in an undergraduate degree such as the one you did? Can you see someone now having the same relationship with rapid prototyping that you had with Fran Hegarty and performance?

Jane: Very much so. This is anecdotal, but when you find out where the rapid prototyping machine is located in the art schools and the university system, and which staff control access to the machine, it is predominantly in engineering with technicians controlling the machine, or it is in design, and there is still a split between art and design. So if you find a rapid prototyping machine you need to ask: Who has access to it? There may be interesting projects going on, but they are mainly the work of designers, because it is the design department that runs the machine, not the art department. That is a problem. Some art departments also shy away from these 'design' technologies and say 'we really want this department to stay very traditional, we don't want laser cutters and rapid prototypers'. It seems very strange to me, to assume that one would replace the other as opposed to them sitting side by side. While such resistance from the visual art departments continues, there are all sorts of things that won't happen.

Nina: But objects such as the heart, that we discussed earlier, couldn't be made by traditionally taught art techniques such as casting. What kind of art object results?

Jane: I think that things that are rapid prototyped that couldn't be made any other way are compelling to us. The cogs could be made another way, but the heart couldn't. Most of us have a fascination with structures that cannot be made using established processes. I think that the novelty value of rapid prototyped objects is strong because everywhere we look, in our homes, social and work life we are surrounded by objects that could be made in all sorts of ways. We rarely come across an object that could only have been made by rapid prototyping. So when we do come across it there is a qualitative difference about that object. Adrian Bowyer[8] has proposed a project he calls RepRap, which is a rapid prototyper that can self-replicate. You have to start with *one*, and then the first thing you do is make another, and then you could give that one away. The idea of enabling mass customization and domestic manufacturing is radical. RepRap challenges the way that technology rolls out, the way that access to it is typically delayed for lower socio-economic groups. I could imagine that a self-replicating rapid prototyper could get taken up quite quickly, early in the development of rapid prototyping for the mass market, by cultures and countries that didn't get hold of inkjet printing early on, and I suspect that if that happened, different sorts of object would get produced that I can't imagine. I think that is one of the unusual things about rapid prototyping, that it is used by artists in the context, or along the trajectory, of new media art. But new media art remains predominantly screen-based, maybe because it came from independent film, guerilla video, guerilla television, independent video, video art, scratch video or performance, rather than from object making. One of my personal hobbyhorses is that if you look at the new media art exhibition circuit, there is no place for the object unless it is interactive. Objects are only allowed into that rarefied ghetto if they are interactive. So, rapid prototyping is problematic for that community; not problematic in the way that they are thinking about it, but problematic because it is not seen as relevant owing to the focus on interactivity and autonomy. When you reference rapid prototyping within the rapid prototyped object, people's interactions are both intellectual and haptic. It is about the 'being' of the object – its form, and how that form came into being. This is what engages you. I think that is completely at odds with most new media work.

8 Adrian Bowyer is senior lecturer in the Department of Mechanical Engineering at the University of Bath working in the Biomimetics Research Group, <http://people.bath.ac.uk/ensab/>; <http://reprap.org/>.

## Chapter 5

# Not Intelligent by Design

Paul Brown and Phil Husbands

One of the key themes that emerged from the formal investigations of art and aesthetics during the twentieth century was that of the autonomous artwork. The goal of an artwork that was not just self-referential but also self-creating found renewed vigour in the work of the systems and conceptual artists and especially those who were early adopters of the, then new, technology of artificial intelligence (AI).

A key problem is that of signature: at what point can we claim that an artwork has its own distinct signature? Co-author Paul Brown's own work in this area began in the 1960s with an early, and in retrospect, naive assumption. At that time art was still based on the concept of engagement with the materiality of the medium. He suggested that using a symbolic language to initiate a process would distance him far enough from the output of that process for it to have the potential of developing its own intrinsic qualities including a unique signature. By the 1990s it had become obvious that this approach had failed. Complementary research in many fields had demonstrated that the signatures of life were robust and strongly relativistic. The myriad bonds that define a signature are embedded in even the simplest symbol system and any attempt to create autonomy by formal construction is unlikely to succeed.

During this same period a group of biologically inspired computational methods were revisited after several decades of neglect; evolutionary, adaptive and learning systems suggested a 'bottom-up' approach to the problem. If it is not possible to design an autonomous agency then can we instead make a system that evolves, learns for itself and eventually has the potential to display autonomy as an emergent property?

The DrawBots project is an attempt to apply these computational methods to the problem of artistic autonomy. It is an example of a *strong* art–science collaboration where all the disciplines involved have a significant investment in the project and its themes. This chapter describes the DrawBots project and its history and discusses its successes and failures to date. It speculates on the potential conclusions of the project and also on its relationship to the larger world of ideas.

## Art that makes itself

It is recounted that the late-nineteenth-century Parisian art dealer Ambroise Vollard during the 96th sitting for his portrait by Paul Cezanne (1839–1906) asked the artist how many more sittings might be needed? Cezanne admonished him, 'If I make one incorrect brush stroke I may have to start the whole painting over again from that mark.' Although this encounter is possibly apocryphal it nevertheless highlights the artist's interest in the formal mechanisms of the craft of painting and helps distinguish him from his contemporaries who were members of the impressionist movement. Claude Monet (1840–1926), for example, revels in his intuitive facility with brush and paint and achieves a close engagement with his subject. By contrast, Cezanne stands apart from his subject, allowing his intellect to consider and govern every move. It is possible that Cezanne's lack of native facility with paint, witnessed by his earlier work, was one reason he adopted this more thoughtful and analytical methodology.

By the late nineteenth century photography had appropriated the role of creating likenesses that had previously been the preserve of painting. This enabled artists such as Cezanne and Georges-Pierre Seurat (1859–1891) to move on from the intuitive, impressionistic representation of the world typical of their contemporaries to begin a more analytical exploration of the relationship between the canvas (the 2-dimensional representation) and the real world (the 3-dimensional scene represented). Their ideas were contemporaneous with, and complementary to, those of the American philosopher Charles Sanders Peirce (1839–1914) and the Swiss linguist Ferdinand de Saussure (1857–1913). Both Peirce and Saussure were developing formal methods for the analysis of systems of communication via an investigation of signs and the relationship between the 'signifier' and the 'signified' that became known as semiotics (Peirce) or semiology (Saussure).

The work and ideas of these artists and philosophers and their contemporaries had a profound effect on the intellectual climate of the nascent twentieth century. In the context of the visual arts – the context of this chapter – they engendered an intense and revolutionary period that lasted into the second half of the century and examined both the purposes and methods of visual production and communication. Experimentation was key to this new spirit and the period culminated in artworks and critical theories that simultaneously questioned and undermined many of the assumptions that were held dear by artists of earlier generations.

By the 1960s a number of key critical concepts had emerged and these included the idea that the process – and not the ensuing object – was the key element of the artwork. Also, the role of art as an intellectual, rather than an emotional, pursuit was emphasized and, in particular, the *idea* or *intention* – the conceptual foundation – of the work was considered paramount.[1] Two international art movements emerged in this period that epitomized these ideas: systems art and conceptual art. Amongst

1 L.R. Lippard, *Six Years: The Dematerialization of the Art Object from 1966 to 1972* (Berkeley: University of California Press, 1973, 1997).

the ideas then current (and not by any means the exclusive domain of the systems and conceptual movements) were autonomy and signature.[2] Mitchell Whitelaw has addressed the origins of these concepts in twentieth-century art, for example in the work of Kasimir Malevich (1878–1935) and Paul Klee (1879–1940).[3] Many artists describe how the artwork itself, during its construction, takes over the creative process and especially how the work itself (and not the artist) dictates the point at which it may be considered complete. This kind of relationship, between the artist and their medium is also implicit in what we know of Cezanne's methodology as indicated above. By the 1960s many artists were engaging explicitly with these ideas. They were attempting to attenuate personality by using industrial materials and methods to remove the human touch and they were adopting formal, structured content and methods that were considered both universal and personality free.

The many influences on this generation of artists included: analytical philosophy; systems theory; artificial intelligence, communications theory; cellular automata (early artificial life); unpredictable deterministic systems (early chaos theory); formal grammars; learning systems and more. Many of these influences were auspiced by a growing awareness of the work of Norbert Wiener (1894–1964) and William Ross-Ashby (1903–1972).[4] Wiener's *Cybernetics*,[5] which first introduced the subject to a wider audience, is subtitled 'the study of control and communication in the animal and the machine' and contributed significantly to a reassessment of the human condition. This finally revoked the Renaissance-inspired view of a humancentric universe – the first-person-singular, perspectival view of the world – and replaced it with one where humans were on a level with other forms of life and even with their machines. It is possible to see that the work of the Cubists some 50 years before – which emerges directly from Cezanne's experiments – was an early progenitor of this heterarchical and multi-perspective world view.

Human superstitions, religion and egocentric concepts of self and importance were, at best, illusionary and human influence was largely peripheral to the workings of the universe. It is perhaps worth noting that humans were not relegated to a position of total inconsequence! George Spencer Brown, a contemporary British analytical philosopher, suggested that humans (and other possible alien life forms) are a mechanism by which the universe is able to perceive itself.[6] This concept continues today in the anthropic principle, which is an essential component of many universe cosmologies like, for example, string theory, where it serves to distinguish

2 At the time of writing the co-authors, together with Margaret Boden and a number of art historians, are developing a research project that will use formal and computational methods to examine the concept of artistic signature.

3 M. Whitelaw, 'The Abstract Organism: Towards a Prehistory for A-Life Art', *Leonardo*, 34/4 (2001): 345–8.

4 W.R. Ashby, *Introduction to Cybernetics* (London: Chapman and Hall, 1956).

5 N. Wiener, *Cybernetics: Or the Control and Communication in the Animal and the Machine* (Cambridge, MA: MIT Press. 1948).

6 G.S. Brown, *Laws of Form* (London: Allen and Unwin, 1969).

this universe from others and, especially, to reason why the fundamental constants that govern this universe have the values that they have. It is also interesting to note, as an aside, that Spencer Brown's work influenced the Chilean biologists Humberto Maturana and Francisco Varela and their development of the concept of autopoiesis.[7]

Back in the slightly less rarefied visual art world of the 1960s these emerging ideas had an equal impact. They reinforced the search for an art that emerges from universal processes rather than from personal fetishes and illusions of self. In the ensuing dialogue a key concept emerged – that of signature.

However, before looking at signature, it is worth examining the work of three pioneers who addressed the influence of cybernetics and what has become known as the computational paradigm and whose work also embeds a claim for autonomy.

Nicolas Schöffer (1912–1992) formulated his idea of a kinetic art that was not only active and reactive, like the work of his contemporaries, but also autonomous and proactive, in Paris in the 1950s. He developed sculptural concepts that he termed spatiodynamism (1948), luminodynamism (1957) and chronodynamism (1959) and was influenced by the new ideas that had been popularized by Wiener and Ross Ashby. His CYSP 1 (1956) is accepted as the first autonomous cybernetic sculpture. Its name is formed from CYbernetic SPatiodynamism 1. It was controlled by an 'electronic brain' (almost certainly an analogue circuit) that was provided by the Dutch electronics company Philips. In addition to its internal movement CYSP 1 was mounted on a mobile base that contained the actuators and control system. Photosensitive cells and a microphone sampled variations in colour, light and sound and so it was:

> excited by the colour blue, which means that it moves forward, retreats or makes a quick turn, and makes its plates turn fast; it becomes calm with red, but at the same time it is excited by silence and calmed by noise. It is also excited in the dark and becomes calm in intense light.

On its second outing CYSP 1 performed with Maurice Béjart's ballet, on the roof of Le Corbusier's Cité Radieuse, as part of the Avant-Garde Art Festival, held in Marseille. Schöffer said of this work:

> Spatiodynamic sculpture, for the first time, makes it possible to replace man with a work of abstract art, acting on its own initiative, which introduces into the show world a new being whose behaviour and career are capable of ample developments.[8]

---

7 L.H. Kauffman and F.J. Varela, 'Form Dynamics', *Journal of Social Biological Structures*, 3 (1980): 171–206.

8 N. Schöffer, quoted from <http://www.olats.org/schoffer/cyspe.htm>. (All URLs current at the time of writing.)

Edward Ihnatowicz (1926–1988) described himself as a Cybernetic Sculptor.[9] His *Sound Activated Mobile* (SAM) consisted of four parabolic reflectors, shaped like the petals of a flower, on an articulating neck. Each reflector focused sound on its own microphone and an analogue circuit could then compare inputs and operate hydraulics that positioned the flower so it pointed towards the dominant sound. SAM would track moving sounds and gave visitors the eerie feeling that they were being observed. Not long afterwards, Ihnatowicz was commissioned by Philips to create the *Senster* for their Evoluon science centre in Eindhoven. It was a large (4m) and ambitious minicomputer-controlled interactive sculpture that responded to sound and movement. It was exhibited from 1970 to 1974, at which point it was dismantled owing to high maintenance costs. Its behaviour was exceptionally life-like[10] and Ihnatowicz was an early proponent of a 'bottom-up' approach to artificial intelligence or what we would now call artificial life. He was inspired by his reading of the developmental psychologist Jean Piaget to suggest that machines would never attain intelligence until they learned to interact with their environment.[11] In recent years he has been widely acknowledged in the scientific world as an early pioneer of what would later be called artificial life or Alife.

Harold Cohen (born 1928) is a well-established artist who represented Britain with his brother Bernard (who later became Slade Professor) at the 1966 Venice Biennale. In 1969 he began working at the University of California at San Diego (UCSD) where he became interested in computers and programming. From 1971 he was involved in the AI Laboratory at Stanford University where Edward Feigenbaum was developing expert systems. These systems get around a major problem in classical, top-down, disembodied AI research – the problem of context (see below). The human mind has an amazing facility to apply quickly a mass of contextual information to the cognition of ambiguities common in speech and other forms of inter-human communication. Even high-speed modern computers with their linear processing structures cannot compete. Feigenbaum was one of a number of researchers in the late 1960s and early 1970s who suggested that this could be overcome by limiting the area of intelligence to small, well-defined knowledge bases where ambiguities could be reduced sufficiently to enable the contextual cross-referencing to be resolved. Researchers at the Stamford lab developed many valuable expert systems, such as Mycin which was used to diagnose infectious diseases and prescribe antimicrobial therapy. As a guest scholar and artist-in-residence from 1971 to 1973, Cohen began to develop an expert system he called AARON. He continues to work on it and jokes that it is the oldest piece of software in continuous development. AARON is a classical top-

9 A. Zivanovic maintains a comprehensive website on Ihnatowicz' work – see <http://www.senster.com>.

10 A. Zivanovic, 'The Technologies of Edward Ihnatowicz', in P. Brown, C. Gere, N. Lambert and C. Mason (eds), *White Heat Cold Logic: British Computer Art 1960–1980* (Cambridge, MA: MIT Press, 2008), 95–110.

11 P. Brown, private conversation with Edward Ihnatowicz (mid 1970s).

down AI package. It contains an internal database and a set of rules that enables it to interpret its knowledge base to produce sophisticated and unique drawings. Although Cohen is interested in investigating issues to do with cognition and drawing in general, his major achievement has been the externalization and codification of his own drawing and cognitive abilities. AARON produces 100 per cent genuine and original Cohen artworks without the need for the human artist's intervention.[12]

## Signature

Recently, in June 2008, someone paid $86 million for Francis Bacon's *Triptych 1976* – the latest record for a work of contemporary art. It is improbable but possible that they bought the painting because it will look good in their corporate or domestic accommodation. It is far more likely that they bought it to deposit in a secure vault for a few years so that they can then sell it on at a good profit. It was described by the vendors as 'totemic' and is believed to be the last remaining major work by the artist that is in private hands. But, really it is just a bit of old fabric with some pigment smeared on it. If we ignore the economic indicators – like recession and inflation – that traditionally favour the art investment market, there are only two good reasons that somebody had $86 million dollars worth of confidence in this investment. Firstly, it is rare – possibly the last remaining piece of old cloth by Bacon that will ever appear on the open market. But more importantly – that bit of paint-smeared fabric bears the attribute of the unique signature style of an artist who is considered important and, even better, who is dead and so incapable of making any more.

The revolution against signature back in the 1960s had two roots. One, described in the previous section, was the challenge of creating a work that existed as a 'pure' manifestation of an idea that was unsullied by the personality (beliefs, prejudices, opinions, attitudes, biases, etc.) of their creator(s). There was also a reaction by artists against the commercial artworld's economic exploitation of their work and this was also related to the artists' rejection of the galleries' focus on the unique (i.e. signed) object at a time when the artists themselves were increasingly concerned with process – the kind of exploitation illustrated by the paragraph above. Co-author Paul Brown was personally involved in this dialogue and although it represents a fairly simplistic overview of what was a more complex situation, it is beyond the scope of this chapter to describe it in further detail. Nevertheless, the rejection of signature – and signature style – was mainly composed of aspects of economic subversion and intellectual/conceptual challenge. This chapter is concerned with the latter.

Signature is not, of course, just – if it ever was – the unique autograph of the artist. Signature is implicit in the artist's choice of subject and medium, and within salient features embedded, often unconsciously, in the execution of that medium.

12 H. Cohen, 'Reconfiguring', in P. Brown et al. (eds), *White Heat Cold Logic*.

In the world of oil painting, for example, the artist's choice of content, minor figurative features, stylistic flourishes, composition, representation of content, preparation of substrate, make of paint, colour palette, type of brushes, the way that paint is mixed and applied and so on, may all contribute salient features that can be identified by a professional assessor who can then use them to make an authoritative attribution of authorship.

In the late 1960s I (Paul Brown) believed that I would be able to make unsigned artworks by using a computer system that I programmed. Prior to this I had first produced flat, geometric paintings using masking tape, liquid acrylic paint and broad soft brushes that enabled me to create an anonymous 'industrial' finish. Then for several years I was artistic director of a lightshow called 'Nova Express'. [Figure 5.1] We were successful and played with many of the major bands of the

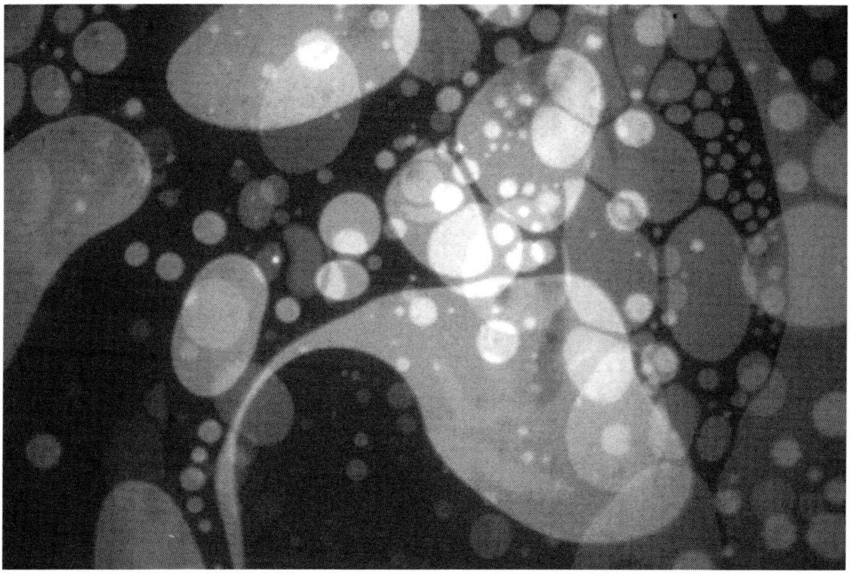

**Figure 5.1 Still image from the 'Nova Express' lightshow**
*Source*: Paul Brown with Jim MacRitchie and Les Parker, c. 1968.

time, such as Pink Floyd, The Who, Nice and Canned Heat, as well as taking part in more in-depth collaborations with leading experimental arts groups including Meredith Monk and the House Company, Electronica Musica Viva and The Welfare State. More importantly, we had sufficient income to invest in equipment and so were able to experiment with a wide range of projection technologies both in rehearsal and performance. The experience of working live with two other operators using both random and structured techniques to integrate our projections with the performers on stage had a major influence on me. I began to see that art could be an ephemeral, a less precious and, significantly, an uncontrolled experience. Back in my studio I could drop some coloured ink into water and watch this process of mixing, which was maybe only 2cm in diameter, projected up to several metres across. The heat of the projection lamp created turbulent flurries and the resulting time-based artwork was an intricate visual fractal – an immersive and absorbing experience. Apart from gently squeezing the ink dropper, I had not really created this event; I didn't try to shape or control it at all. I had nowhere near the amount of control I would have exercised if I had been painting a canvas or carving a sculpture. Factors such as heat, gravity and turbulence were completely beyond my control – the work existed as a visualization of a physical event in contrast to a deliberately created aesthetic object: I was watching the laws of physics and chemistry working 'live' on the screen. Furthermore, I could build a machine to squeeze the dropper! My longstanding interest in autonomy began. Around the same time I also began to experiment with chemically-modified photographs [Plate 5.1] and electrostatics [Figure 5.2].

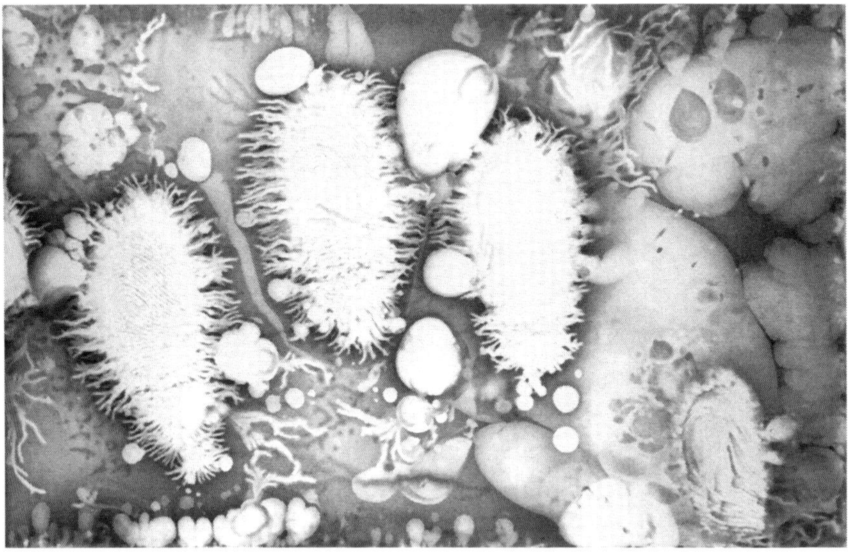

**Figure 5.2 Electrograph of hand**

*Source*: Paul Brown, c. 1974.

I had discovered computers at the Cybernetic Serendipity exhibition at the ICA in 1968[13] and by 1974 was using them exclusively in my work. The computer was a machine and produced images using an offline Calcomp pen plotter – so the entire process was automated and I didn't have to engage physically with the work

13 J. Reichardt, 'In the Beginning', in P. Brown et al. (eds), *White Heat Cold Logic*.

throughout its production. Furthermore by employing a simple formal symbolic programming language like FORTRAN,[14] I thought I should be able to distance myself further from the work. By utilizing very simple drawing 'primitives' and distributing these about the image space, first by using random procedures and later using the agency of cellular automata, I planned to produce work that would have the potential to develop a unique and autonomous signature and that this would be significantly different to my own. I believed that using a symbolic language could initiate a process that would enable me to distance myself far enough from that process and its outputs for it to have the potential to develop its own intrinsic qualities including a unique signature.

It retrospect it was an over-optimistic expectation but at the time it seemed reasonable and led directly to a couple of decades of interesting and productive engagement with computer systems, computational theory, artificial intelligence, artificial life, and so on. Some of the works I produced are shown in Figures 5.3 and 5.4. More are illustrated in Catherine Mason's insightful history of the computer arts in the UK and on my website.[15] However by the early 1990s, after 20 years following this particular avenue, it became obvious that – however interesting or valuable the work I had produced – the fundamental aim of autonomy had not been achieved: in that respect the work was a cul-de-sac. The artworks that had been created were clearly signed with my own name, inasmuch as they evinced a particular style and mode of practice with which I could be identified.

During the second half of the twentieth century we learned a great deal about the signatures of life, their codes and manifestations. Signatures, like life itself, are extremely robust and not easy to ignore, disguise or overcome. Research in many fields has demonstrated that they are strongly relativistic. The myriad bonds that define a signature are transmitted by even the simplest symbol system and for this reason any attempt to create autonomy by formal construction, as I had attempted, is unlikely to succeed.

I was not too disappointed by my failure. The work I had made was interesting and had further potential. The quest for autonomy was put to one side. It re-emerged in 2000 when, as the recipient of an Australia Council New Media Arts Fellowship, I spent a year at the Centre for Computational Neuroscience and Robotics (CCNR) at the University of Sussex in the UK. Here I learned about many exciting new developments in artificial life and these inspired me to readdress the question of autonomous artworks.

The following section introduces the biologically inspired form of AI that underpins the DrawBots project. By explaining how the methodology used – that of evolutionary robotics – fits into the history of AI, it gives important context to the work. The section also discusses relationships between art and science in the CCNR.

---

14 FORTRAN or FORmula TRANslation was an early 'high-level' language devised for mathematical and scientific applications and it was the first programming language I learned.

15 C. Mason, *A Computer in the Art Room: The Origins of British Computer Arts 1950–80* (Norfolk: JJG, 2008); <http://www.paul-brown.com/>.

**Figure 5.3 *LifeMods*, plotter drawing**

*Source*: Paul Brown, c. 1977.

**Figure 5.4** ***36 Knots for Fu Hsi*, microfilm plot**
*Source*: Paul Brown, c. 1977.

## GOFAI, NEWFAI, top-down and bottom-up

In 1969 Marvin Minsky and Seymour Papert published a book that proved that a particular kind of early artificial neural network (single-layer perceptrons)[16] were fundamentally flawed in their ability to learn and could never be trained to recognize many important classes of input patterns.[17] Minsky and Papert conjectured that a similar result would hold for the general class of perceptron

16 F. Rosenblatt, 'The Perceptron: A Probabilistic Model for Information Storage and Organization in the Brain', *Cornell Aeronautical Laboratory, Psychological Review*, 65/6 (1958): 386–408.

17 M.L. Minsky and S.A. Papert, *Perceptrons* (Cambridge, MA: MIT Press, 1969).

systems, including those with more layers. The implication was that this kind of brain-inspired learning device was a dead end. It was no coincidence that at the time Minsky was spearheading an alternative approach to machine intelligence – one based around computer programs that explicitly manipulated symbols representing elements of the problem at hand. In fact, the conjecture about the general perceptron result turned out to be completely wrong. Just three years after Minsky and Papert published their book, Grossberg introduced a class of neural network that overcame the limitations of single-layer perceptrons,[18] and indeed two years before the book appeared Amari had proposed a solution to the problem[19] that was later shown to be correct. But Minsky was a powerful advocate of the symbolic approach and an equally powerful detractor of neural networks; the damage had been done and funding for research in learning machines went into steep decline. For the next 20 years the symbol-processing approach dominated AI. But this had not always been the case and today biologically inspired approaches based on self-organizing adaptive systems are again centre stage. To understand the significance of this switch, and to appreciate the differences between the two major approaches, some history is in order.

The history of machine intelligence is longer than most people realize – stretching back at least to Hobbes[20] – and certainly too long to summarize here,[21] but it is generally acknowledged that it first started to gain serious momentum in the 1940s. In Britain and the US, groups of scientists came together, intent on understanding the general principles underlying behaviour in animals and machines. Their interdisciplinary approaches mixed ideas from biology, engineering and mathematics. The mathematician at the centre of the American group, Norbert Wiener, coined the term 'cybernetics' to describe the enterprise.[22] For a while cybernetics flourished and produced many important ideas and techniques as well as being centrally involved in the development of computers. Foremost amongst the scientific groupings were the Macy conferences in the US[23] and the Ratio Club in Britain.[24] Cybernetic approaches to machine intelligence

18 S. Grossberg, 'Contour Enhancement, Short-term Memory, and Constancies in Reverberating Neural Networks', *Studies in Applied Mathematics*, 52 (1973): 213–57.

19 S. Amari, 'Theory of Adaptive Pattern Classifiers', *IEEE Transactions in Electronic Computers*, 16 (1967): 299–307.

20 T. Hobbes, *Leviathan* (London: Andrew Crooke, 1651).

21 See, for example, M.A. Boden, *Mind as Machine: A History of Cognitive Science* (Oxford: Oxford University Press, 2006) and P. Husbands, O. Holland and M. Wheeler (eds), *The Mechanical Mind in History* (Cambridge, MA: MIT Press, 2008) for much fuller coverage.

22 Wiener, *Cybernetics*.

23 S. Heims, *Constructing a Social Science for Postwar America: The Cybernetics Group, 1946–1953* (Cambridge, MA: MIT Press, 1991).

24 P. Husbands and O. Holland, 'The Ratio Club: A Hub of British Cybernetics', in P. Husbands, O. Holland and M. Wheeler (eds), *The Mechanical Mind in History* (Cambridge, MA: MIT Press, 2008), pp. 91–148.

were strongly biologically motivated with achievements including the introduction of the first artificial neural networks by McCulloch and Pitts,[25] the development of ideas about learning machines and evolutionary systems by Turing,[26] Ashby's general theories of adaptation,[27] and Walter's development of the first autonomous mobile robots,[28] controlled by simple electronic nervous systems, to name but a few. Cybernetics emphasized adaptive 'bottom-up' approaches, often based on the interactions between simple neuron-inspired elements, in which systems learned and adapted, rather than having their target behaviours explicitly designed. This led to Selfridge's breakthrough Pandemonium system in the mid 1950s which learned to recognize visual patterns, including alphanumeric characters.[29] The system employed a layered network of processing units that operated in parallel and made use of explicit feature detectors that only responded to certain visual stimuli. At about the same time Rosenblatt developed the perceptron learning system discussed earlier. As mentioned in the previous section, the influence of cybernetics soon spread beyond science into the arts.

In 1956 John McCarthy and Marvin Minsky organized a long workshop at Dartmouth College[30] to develop new directions in what they termed *artificial intelligence*. (The new label was introduced at a meeting they had organized with Oliver Selfridge the previous year.)[31] In particular, McCarthy proposed using newly available digital computers to explore the radical psychologist Kenneth Craik's conception of intelligence which highlighted the use of internal models of external reality,[32] emphasizing the power of symbolic manipulation of such models. At the meeting, Allen Newell and Herbert Simon, influenced by aspects of Selfridge's work, demonstrated a symbolic reasoning program that was able to solve problems in mathematics. These impressive early results helped drive the rise of logic-based, symbol-manipulating computer programs in the study of machine intelligence. This more abstract, software-bound paradigm came to dominate the field and pulled it away from its biologically inspired origins. For a while the term artificial intelligence, or AI, was exclusively associated with this

25 W.S. and W. Pitts, 'A Logical Calculus of the Ideas Immanent in Nervous Activity', *Bulletin of Mathematical Biophysics*, 5 (1943): 115–33.

26 A.M. Turing, 'Computing Machinery and Intelligence', *Mind*, 59 (1950): 433–60.

27 W.R. Ashby, 'Adaptiveness and Equilibrium', *Journal of Mental Science*, 86 (1940): 478 and W.R. Ashby, *Design for a Brain* (London: Chapman and Hall, 1952).

28 W. Grey Walter, 'An Imitation of Life', *Scientific American*, 182/5 (1950): 42–5.

29 O.G. Selfridge, 'Pandemonium: A Paradigm for Learning', in D. Blake and A. Uttley (eds), *The Mechanisation of Thought Processes. Volume 10 of National Physical Laboratory Symposia* (London: HMSO, 1959), pp. 511–29.

30 J. McCarthy, M. Minsky, N. Rochester and C. Shannon, 'A Proposal for the Dartmouth Summer Research Project on Artificial Intelligence' (1955), <http://www-formal.stanford.edu/jmc/history/dartmouth/dartmouth.html>.

31 The Western Joint Computer Conference, Los Angeles, 1–3 March 1955.

32 K.J.W. Craik, *The Nature of Explanation* (Cambridge: Cambridge University Press, 1943).

style of work. This paradigm, which to some extent harked back to the older ideas of George Boole and Gottfried Leibniz, also served as a new kind of abstract model of human reasoning, becoming very influential in psychology and, later, in cognitive science.

This style of AI, which later came to be known as GOFAI (Good Old Fashioned AI) employed a top-down approach. The problem of producing intelligent behaviour was broken down into stages for which explicit solutions were designed. The idea was to program-in intelligence, rather than making use of learning and adaptation as happens in nature. The whole paradigm is neatly illustrated by the famous early AI robot, Shakey, developed at the Stanford Research Institute (SRI). The robot accepted goals from the user, planned how to achieve them and then executed those plans via intermediate-level actions (ILAs) implemented as predefined routines.[33] The ILAs were translated into complex routines of low-level actions that had some error detection and correction capabilities and which dealt directly with the robot hardware. The overall processing loop had at its heart the sequence of operations shown in Figure 5.5. Here, robot intelligence is functionally decomposed into a strict pipeline of operations. Central to this view of intelligence is an internal model of the world which must be built, maintained and constantly referred to in order to decide what to do next. In Shakey's case, as in much AI at the time, the world model was defined in terms of formal symbolic logic: it consisted of a collection of predicate calculus statements.

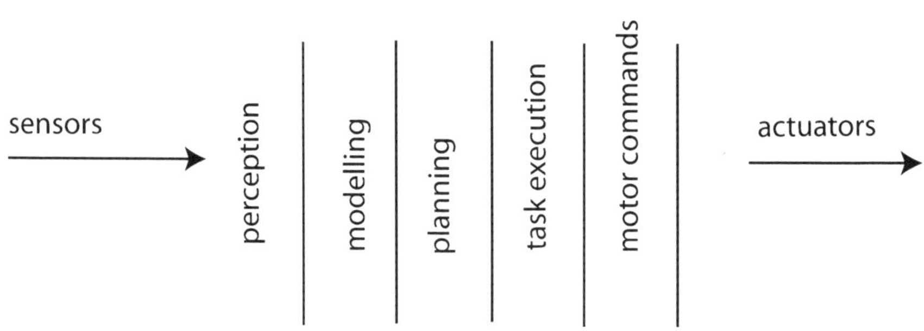


**Figure 5.5 Pipeline of functionally decomposed processing used in much AI robotics**

*Source*: After Brooks, 1986.

Shakey's reasoning processes could be computationally very expensive, even in the carefully constructed environments in which it operated (these consisted mainly of large coloured blocks of various regular shapes and sizes). However, the

33 N.J. Nilsson (ed.) *Shakey the Robot*, Technical Note 323 (Menlo Park CA: AI Center, SRI International, 1984).

general approach was very influential and it dominated for more than a decade, during which time the individual functions, as shown in Figure 5.5, tended to become separate specialisms which started to lose contact with each other, and the importance of embodiment and interaction with an environment were lost. Shakey, and similar robots, would often take hours to complete a single task and were reliant on the programmer modelling the world and the ways in which it could change such that all eventualities were encompassed. By the mid 1980s a number of leading researchers from the main AI robotics centres were becoming more and more disillusioned with the approach.

Hans Moravec, an influential roboticist who had done important work on the Stanford Cart, a project similar in spirit and approach to SRI's Shakey and which ran at about the same time,[34] summed up such feelings:

> For the last fifteen years I have worked with and around mobile robots controlled by large programs that take in a certain amount of data and mull it over for seconds, minutes or even hours. Although their accomplishments are sometimes impressive, they are brittle – if any of many key steps do not work as planned, the entire process is likely to fail beyond recovery.[35]

Moravec goes on to point out how this is in strange contrast to the pioneering work of cyberneticist Grey Walter and the projects that his early mobile robots inspired; these devices' simple sensors were connected to their motors via fairly modest circuits and yet they were able to behave very competently and 'managed to extricate themselves out of many very difficult and confusing situations' without wasting inordinate amounts of time 'thinking'.

By the mid 1980s the GOFAI approach was faltering in many areas of AI, not just robotics. Vastly complex programs would often grind away for hours before reaching a conclusion. As Moravec pointed out, the whole approach was often very brittle, relying on the accuracy and sufficiently broad scope of the knowledge programmed in. Those disillusioned with this state of affairs found a particularly effective voice in Rodney Brooks, who was developing an alternative vision not only of intelligent robotics, but also of the general AI problem. The combative figure of Brooks, along with his team at MIT, became central to a growing band of dissidents who launched a salvo of attacks on the AI mainstream. In a move that conjured up the spirit of cybernetics, the dissidents rejected the assumptions of the establishment, instead regarding the major part of natural intelligence to be closely bound up with the generation of adaptive behaviour in the harsh unforgiving environments most animals inhabit. Hence the investigation of strongly biologically inspired complete embodied autonomous sensorimotor

---

34 H. Moravec, 'The Stanford Cart and The CMU Rover', *Proceedings of the IEEE*, 71/7 (1983): 872–84.

35 H. Moravec, 'Sensing Versus Inferring in Robot Control', Informal Report (1987), <http://www.frc.ri.cmu.edu/~hpm/project.archive/robot.papers/1987/sense.ltx 1987>.

systems – 'artificial creatures' – was seen as the most fruitful way forward, rather than the development of disembodied algorithms for abstract problem solving. The central nervous system was viewed as a fantastically sophisticated control system, not a chess-playing computer.[36] At the heart of Brooks' approach was the idea of behavioural decomposition as opposed to traditional functional decomposition. The overall control architecture involves the coordination of several loosely coupled behaviour generating systems all acting in parallel. Each has access to sensors and actuators and can act as a standalone control system. Each layer was thought of as a *level of competence*, with the simpler competences at the bottom of the vertical decomposition and the more complex ones at the top.[37]

Brooks' work triggered the formation of the so-called 'New AI' movement – which we might refer to as NEWFAI (NEW Fangled AI) – still strong today. With its focus on the development of whole artificial creatures as an important way to deepen our understanding of natural intelligence as well as to provide new directions for the engineering of intelligent machines, this movement has strong links to the pre-AI cybernetic roots of machine intelligence.

The explosion in biologically inspired approaches to robotics and AI that Brooks' work helped to fuel brought forth numerous interesting strands of research, many of which are still very active, and helped to bring AI much closer to biology, particularly neuroscience, than it had been for many years. Although GOFAI did have its successes (e.g. expert systems acting in sufficiently narrow domains), it had practically ignored adaptation and learning for decades and as a consequence, as we have seen, many of its methods suffered from brittleness. As these limitations became more obvious, other biologically inspired areas such as neural networks, adaptive systems, artificial evolution and artificial life all came to the fore. These various currents mingled with the 'New AI' approaches to robotics, spawning new attitudes and directions. The face of AI was radically changed and bottom-up adaptive approaches were once again at the top of the agenda.

One of the new approaches that emerged at this time was that of evolutionary robotics, a field of which I (Phil Husbands) am one of the founders. This is the main methodology used in the DrawBots project, described later.

Alan Turing's paper 'Computing Machinery and Intelligence'[38] is widely regarded as one of the seminal works in artificial intelligence. It is best known for what came to be called the Turing test – a proposal for deciding whether or not a machine is intelligent, involving an interrogator asking questions, via a teletype, to players A and B, one of whom is human, the other a machine. If on average the interrogator cannot tell which player is the machine and which human, the machine might be regarded as intelligent. However, tucked away towards the end of Turing's

36 R.A. Brooks, *Cambrian Intelligence: The Early History of the New AI* (Cambridge, MA: MIT Press, 1999).

37 R.A. Brooks, 'A Robust Layered Control System for a Mobile Robot', *IEEE Journal of Robotics and Automation*, 2/1 (1986): 14–23.

38 Turing, 'Computing Machinery and Intelligence'.

Plate 3.1 Stanza, *Sensity*, 2004–2008. Installation shot on 3D globe, County Hall, London using Pufferfish Globe. Wireless sensors, networked media, generative system, real time data visualization.

**Plate 3.2** Stanza, *Soul*, 2004–2007. Live CCTV and networked media experience on 3D globe. Proposal for Turbine Hall, Tate Modern.

**Plate 4.1** Simulated English oak from an algorithm by Gordon Selley as part of *Decoy*, Jane Prophet, 2001.

**Plate 4.2** Rapid prototyped tree made from edited version of Selley's algorithm with assistance from Adrian Bowyer. Part of *Model Landscapes*, Jane Prophet, 2005.

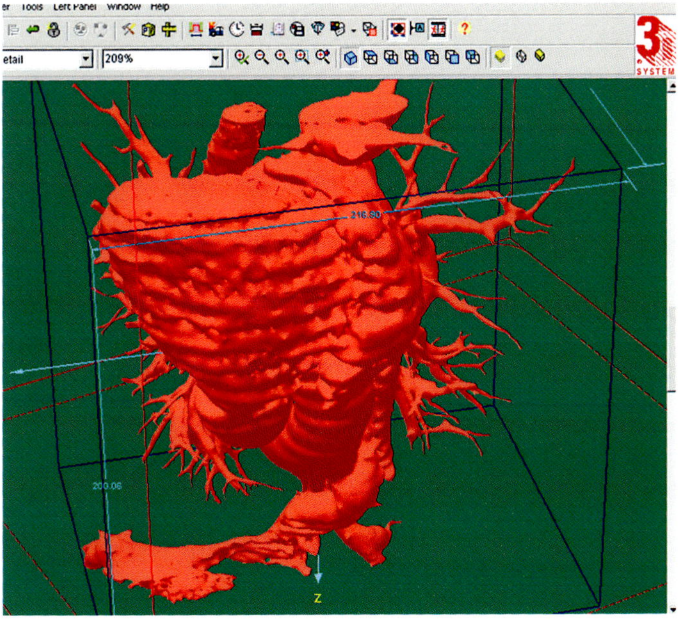


**Plate 4.3** MRI image of healthy heart, visualized using 3D software, Jane Prophet, 2004.

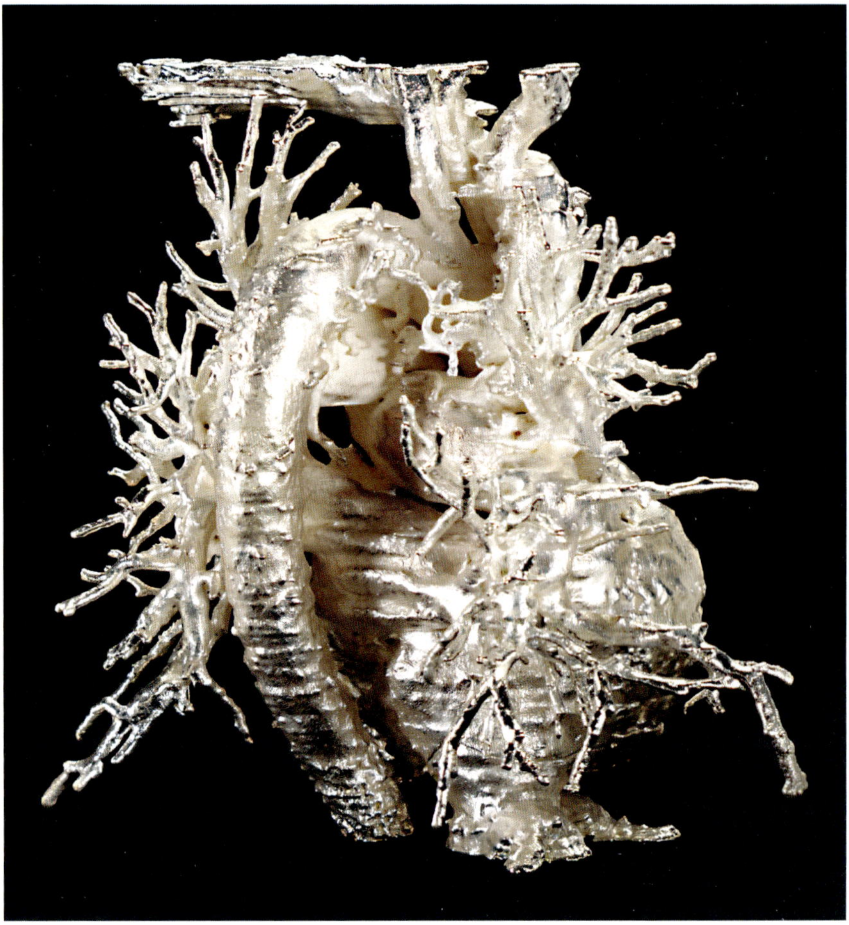

**Plate 4.4** **Silver heart created from 3D MRI data by coating a polymer rapid prototype with a thin layer of copper, followed by a coat of silver, Jane Prophet, 2004.**

**Plate 5.1 Chemically modified photograph, Paul Brown, 1969.**

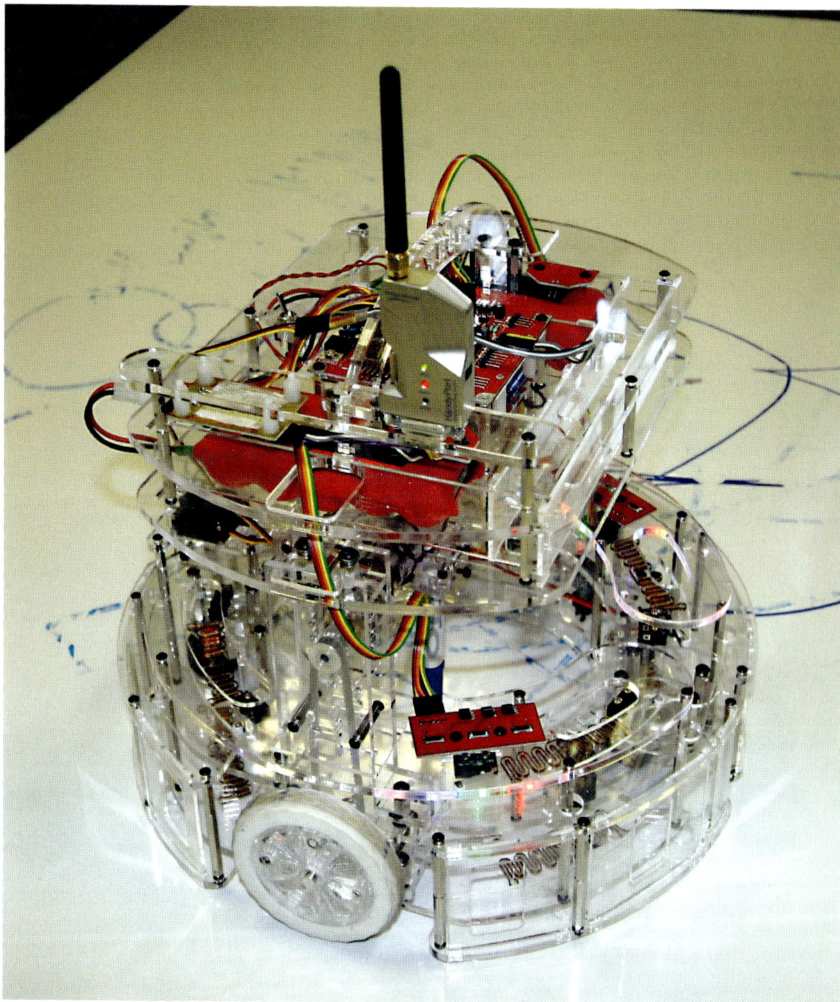

**Plate 5.2 DrawBot V3.0, Bill Bigge, 2007.**

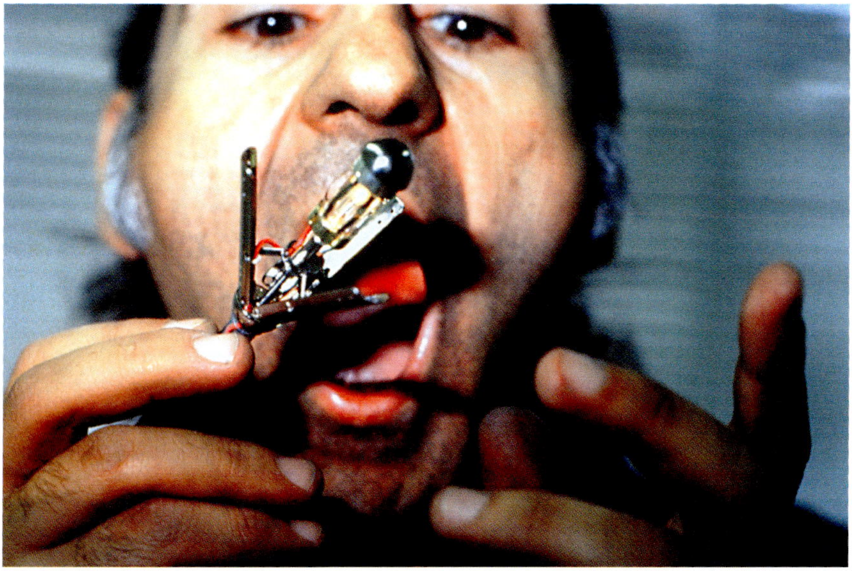

**Plate 6.1** **Stelarc, *Stomach Sculpture*, 1993. Fifth Australian Sculpture Triennale, Melbourne 1993. Photograph by Anthony Figallo.**

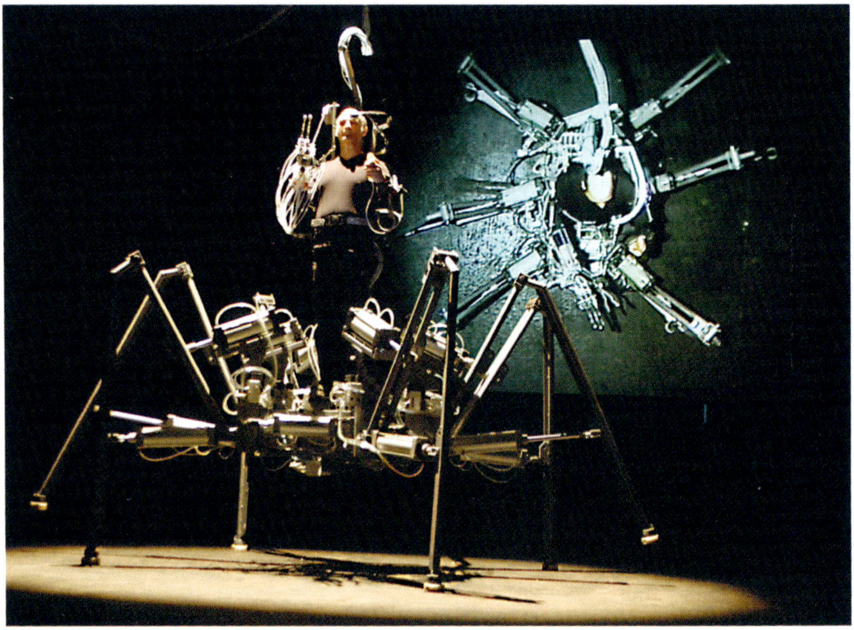

**Plate 6.2** **Stelarc, *Exoskeleton*, 2003. Cankarjev Dom, Ljubljana. Photograph by Igor Skafar.**

**Plate 6.3** **Stelarc, *Blender*, 2005. Teknikunst, Melbourne. Photograph by Stelarc.**

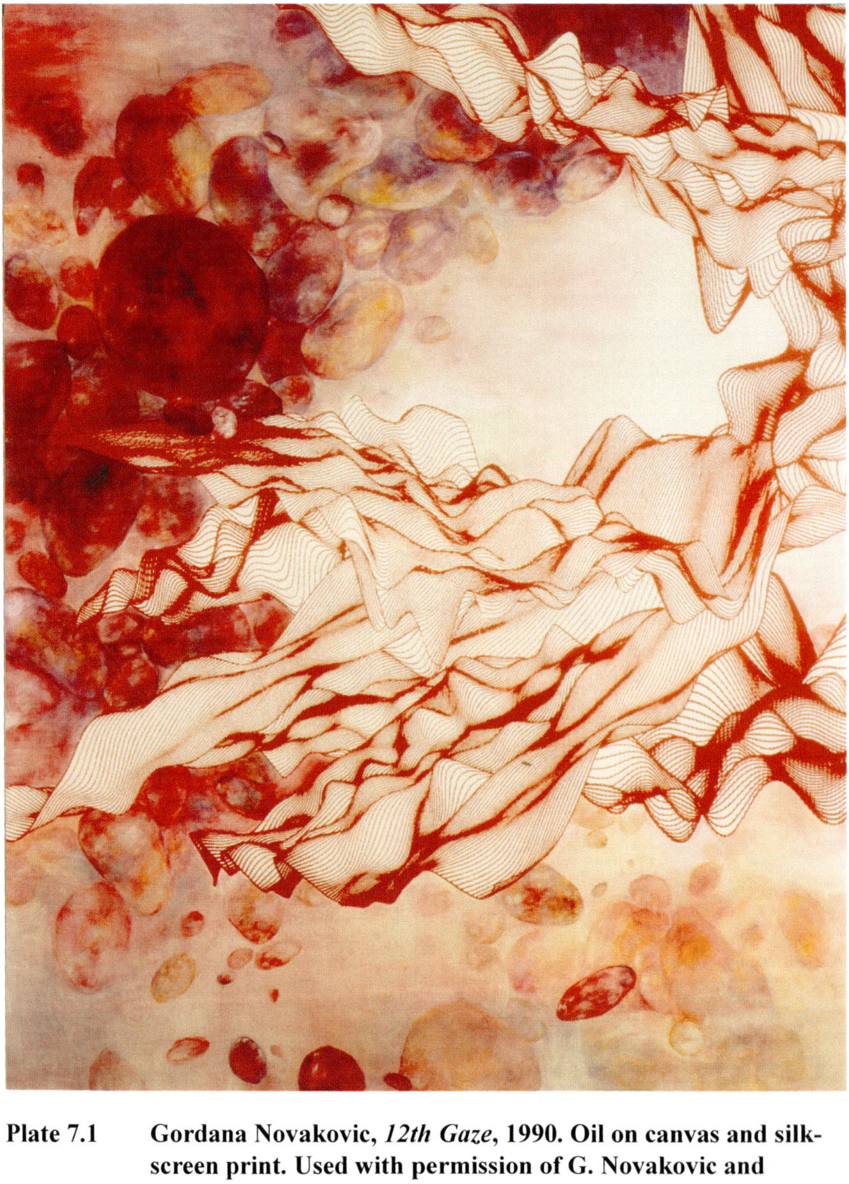

Plate 7.1 Gordana Novakovic, *12th Gaze*, 1990. Oil on canvas and silk-screen print. Used with permission of G. Novakovic and A. Zlatanović.

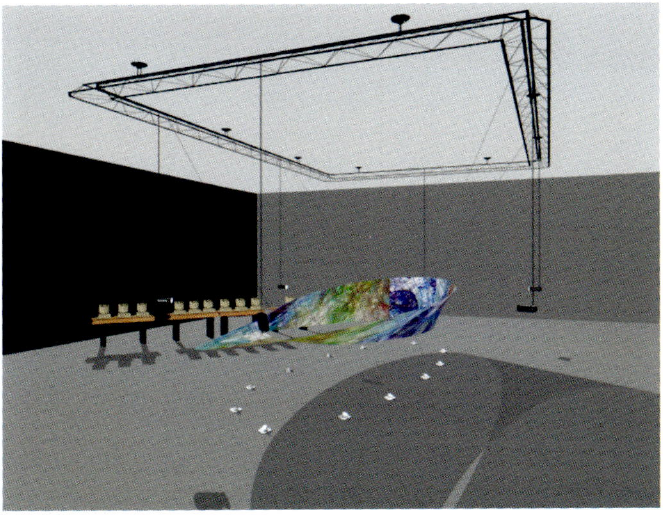

Plate 7.2 A still from the 3D computer animation of Infonoise, showing the suspended Möbius strip and the bank of computers for controlling the installation (2001). The twelve white objects arranged in an oval beneath the Möbius strip represent the proximity sensors used to detect the movements of the participants. Used with permission of G. Novakovic and M. Mandić.

Plate 7.3 A participant about to enter the *Fugue* interactive installation (2006). The lamp above the installation provides a strong infrared light for tracking the participants. The apparatus in front of the enclosure controls the interactive sound and projection system. Still from the video-documentary. Used with permission of G. Novakovic and R. Novakovic.

Plate 7.4 A close-up of a real-time generated image of immune system elements responding to a participant in the *Fugue* interactive installation (2006). Used with permission of G. Novakovic and R. Linz.

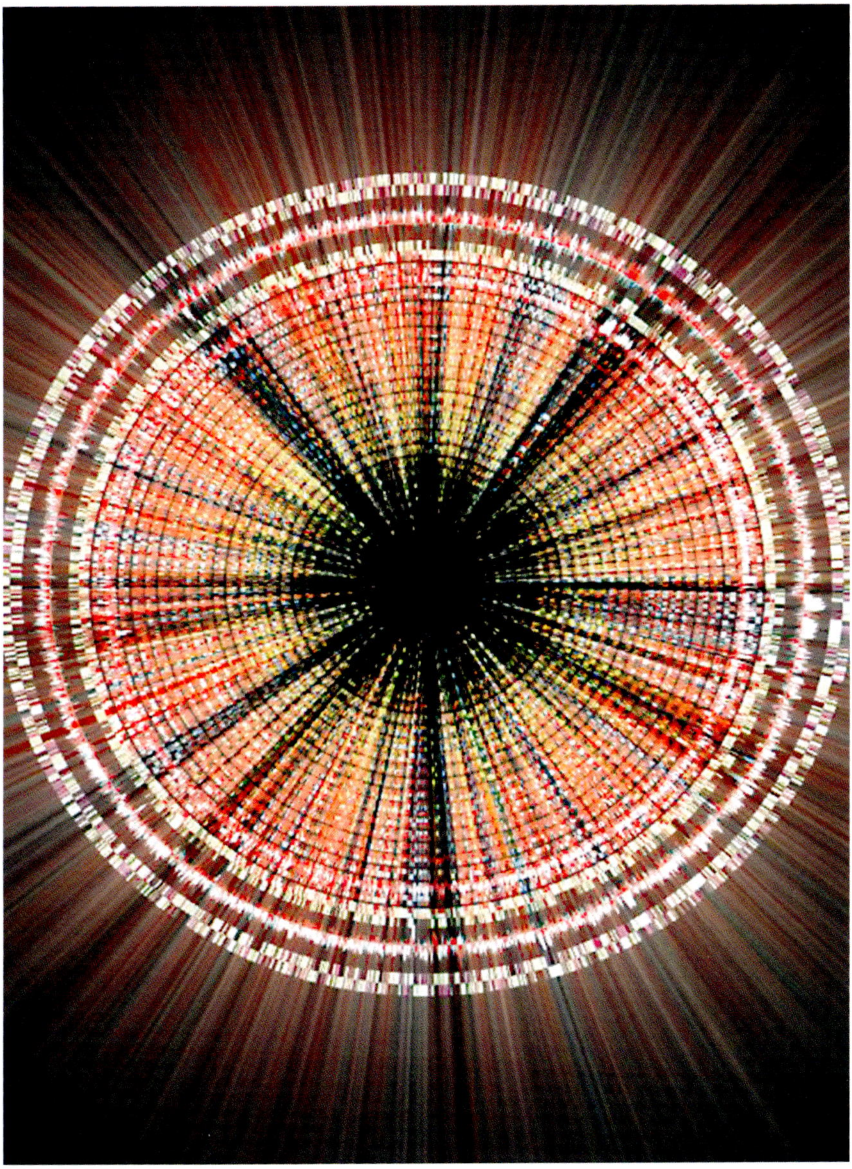

**Plate 8.1** **Elaine Shemilt, *Blueprint for Bacterial Life*, 2008. Still, from animation.**

**Plate 8.2 Elaine Shemilt, *Rings and Rays*, 2007. Still, from animation.**

**Plate 9.1 Paul Sermon, *Headroom*, 2006. Video still.**

**Plate 9.2** **Paul Sermon, *Liberate Your Avatar*, Manchester, 2007. A merged reality performance.**

**Plate 9.3** **Paul Sermon, *Memoryscape*, Taipei, 2006. Visitors explore the augmented memoryscape.**

wide-ranging discussion of issues arising from the test is a far more interesting proposal. He suggests that worthwhile intelligent machines should be adaptive, should learn and develop, but concedes that designing, building and programming such machines by hand is probably completely infeasible. He goes on to sketch an alternative way of creating machines based on an artificial analogue of biological evolution. Each machine would have hereditary material encoding its structure, mutated copies of which would form offspring machines. A selection mechanism would be used to favour better-adapted machines – in this case those that learned to behave most intelligently. Turing proposed that the selection mechanism should largely consist of the experimenter's judgement.

It was not until more than 40 years after their publication that Turing's long forgotten suggestions became reality. Building on the development of principled evolutionary search algorithms by, among others, John Holland,[39] researchers at the Consiglio Nazionale delle Ricerche (CNR) in Rome, Ecole Polytechnique Fédérale de Lausanne (EPFL) in Switzerland, the University of Sussex and Case Western University in Cleveland, Ohio, independently demonstrated methodologies and practical techniques to evolve, rather than design, control systems for primitive intelligent machines. It was out of the spirited milieu of NEWFAI that the field of evolutionary robotics was born in the early 1990s.[40] Initial motivations were similar to Turing's: the hand design of intelligent adaptive machines intended for operation in natural environments is extremely difficult. Would it be possible to wholly or partly automate the process? From the outset most work in this area has involved populations of artificial genomes (lists of characters and numbers) encoding the structure and other properties of artificial neural networks that are used to control autonomous mobile robots required to carry out a particular task or to exhibit some set of behaviours. Other properties of the robot, such as sensor layout or body morphology, may also be under genetic control. The genomes are mutated and interbred, creating new generations of robots according to a Darwinian scheme in which the fittest individuals are most likely to produce offspring. Fitness is measured in terms of how good a robot's behaviour is, according to some evaluation criteria; this is usually automatically measured but may, in accordance with Turing's original proposal, be based on the experimenters' judgement. Work

39 J.H. Holland, *Adaptation in Natural and Artificial Systems* (Ann Arbor: University of Michigan Press, 1975).

40 D.T. Cliff, I. Harvey and P. Husbands, 'Explorations in Evolutionary Robotics', *Adaptive Behaviour*, 2/1 (1993): 71–108; P. Husbands and I. Harvey, 'Evolution versus Design: Controlling Autonomous Mobile Robots', in *Proceedings of the 3rd Annual Conference on Artificial Intelligence, Simulation and Planning in High Autonomy Systems* (Los Alimitos, CA: IEEE Computer Society Press, 1992), pp. 139–46; and D. Floreano and F. Mondada, 'Automatic Creation of an Autonomous Agent: Genetic Evolution of a Neural-network Driven Robot', in D. Cliff, P. Husbands, J. Meyer and S.W. Wilson (eds), *From Animals to Animats III: Proceedings of the Third International Conference on Simulation of Adaptive Behavior* (Cambridge, MA: MIT Press/Bradford Books, 1994), pp. 402–10.

in evolutionary robotics is now carried out in many labs around the world and numerous papers have been published on many aspects of the field.[41]

The key elements of the evolutionary robotics approach are illustrated in Figure 5.6 and are:

- An artificial genetic encoding specifying the robot control systems/body plan/sensor properties, etc., along with a mapping to the target system (the genome and genotype to phenotype mapping, respectively).
- A method for measuring the fitness of the robot behaviours generated from these genotypes (the fitness function).
- A way of applying selection and a set of 'genetic' operators, such as mutation, to produce the next generation from the current (the details of the evolutionary search algorithm).

Potential advantages of this methodology include:

- The ability to explore potentially unconstrained designs that have large numbers of free variables. A *class* of robot systems (to be searched) is defined rather than fully specifying particular robot designs. This means fewer assumptions and constraints are necessary in specifying a viable solution.
- The ability to use the methodology to fine-tune parameters of an already successful design.
- The ability, through the careful design of fitness criteria and selection techniques, to take into account multiple, and potentially conflicting, design criteria and constraints (e.g. efficiency, cost, weight, power consumption, etc.).
- The possibility of developing highly unconventional and minimal designs.

Many different kinds of robot behaviour have been successfully evolved, including various kinds of walking, visually guided behaviours, flying, complex navigation, group behaviours, as well as self-repairing behaviours.[42]

As a bottom-up technique where the solution is not specified in advance, and where many different and unexpected behaviours often emerge, it is clear why the evolutionary robotics methodology is potentially very interesting in relation to the issues of artistic signature as discussed in the previous section.

---

41 S. Nolfi and D. Floreano, *Evolutionary Robotics: The Biology, Intelligence, and Technology of Self-organizing Machines* (Cambridge, MA: MIT Press/Bradford Books, 2000) and D. Floreano, P. Husbands and S. Nolfi, 'Evolutionary Robotics', in B. Siciliano and O. Khatib (eds), *Springer Handbook of Robotics* (Berlin: Springer, 2008), pp. 1423–51.

42 Floreano et al., 'Evolutionary Robotics'.

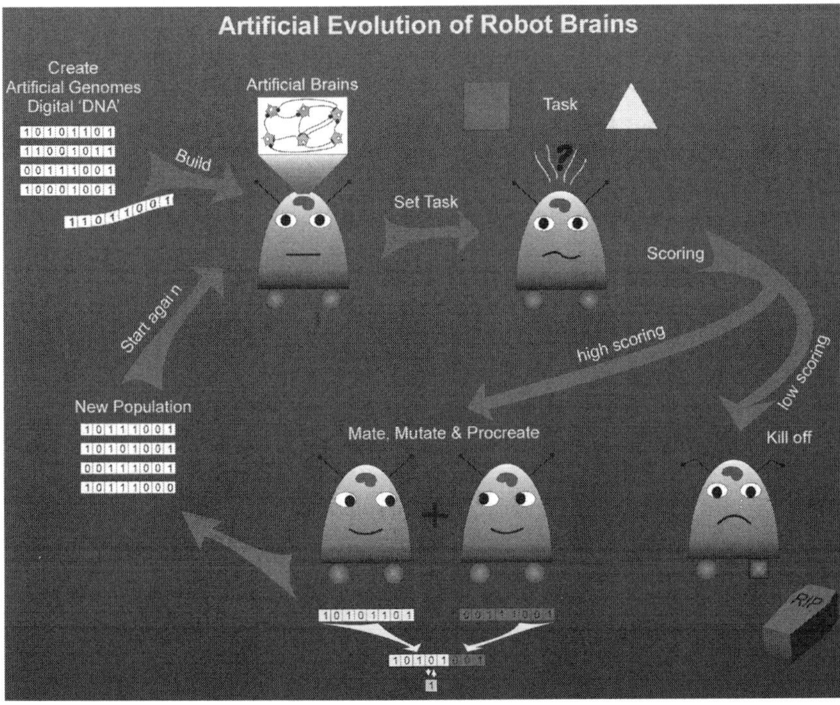


**Figure 5.6 Key elements of the evolutionary robotics approach**

Those of us involved in NEWFAI, like our cybernetic forefathers, often have interdisciplinary backgrounds. I originally trained in physics and mathematics but have always been involved in the arts, mainly as a musician (but also as a published writer of fiction). So it is perhaps no surprise that ever since Michael O'Shea (a prominent Sussex neuroscientist) and I (Phil Husbands) set up The Sussex Centre for Computational Neuroscience and Robotics (CCNR) in 1996 it has collaborated with a variety of artists and hosted several artists in residence including: Stelarc, Paul Brown, John McCormack, Sol Sneltvedt, Rachel Cohen, Anna Dumitriu and Murray McKeich.

We can think of art and science as dealing with different aspects of the same venture: revealing the world.[43] However, in doing this there are a number of ways in which they can interact directly, most of which we have experience of in the CCNR.

The ways in which we perceive and interact with the world are to a large degree determined by our biology. There are various natural biases in the ways we see and hear: our minds and bodies have evolved in such a way that we are set up to respond preferentially to certain kinds of stimuli. Most of this occurs

43 C. Biederman, *Art as the Evolution of Visual Knowledge* (Red Wing: MI, 1948).

at a subconscious level and many of the biases are hidden deep within the huge complexities of our nervous systems. Hence the scientific study of the perception and creation of art might give insights into ancient and important workings of the mind. In such a case art becomes the inspiration and subject matter of science. The reverse of this is equally important, where science provides the inspiration and background material for an artwork. These two areas have resulted in a number of fruitful collaborations based at the CCNR, including the ACE and AHRC-funded Mindscape project involving artists Sol Sneltvedt and Charlie Hooker working with Michael O'Shea and other members of the CCNR on an artwork that gave visual and audio representations of brain dynamics.

A related kind of interaction involves the artist seeing the aesthetic in what the scientist may regard as merely data. An interesting example of this occurred when Paul Brown first had a residency at the CCNR. One of our researchers, Kyran Dale, was presenting some graphical representations of the flight paths of some artificially-evolved virtual insects that he was studying to try to gain insights into the ways that real insect brains work. Paul saw these as rather exquisite drawings, as he explains in the following section on the DrawBots project.

The fundamental preoccupations of the scientists and artists in the CCNR have often overlapped to a considerable degree – for instance, in exploring the ways in which simple interacting adaptive processes can give rise to complex patterns in space and time, or in questioning how we, as embodied intelligent agents, interact with our surroundings. In some cases this leads to another, very direct, kind of relationship between art and science: the development and use of technology and scientific tools for artistic expression. In these cases the boundaries between art and science are considerably blurred as researchers in the group develop and apply biologically inspired adaptive technology to creative domains – for example, in the composition and performance of music, design of sounds or creation of visual artworks. In some cases the boundaries are completely dismantled, as art and science are merged into a single enterprise– for example, in the DrawBots project mentioned above. This kind of activity, in which our artists-in-residence play an inspirational role, has led to a number of our researchers, such as Alice Eldridge, Jon Bird, Bill Bigge and James Mandelis, developing their work in very interesting directions, right on the boundary between art and science. This group – along with the late Drew Gartland-Jones, who played a crucial role in engendering art–science collaborations in the CCNR – also established the very successful Brighton-based art–science forum, Blip.

The main focus of the CCNR – work at the interface between the biological and computational sciences aimed at better understanding natural and artificial adaptive systems – is intrinsically interdisciplinary, with much of the territory uncharted. Perhaps this attracts a certain kind of creative scientist and encourages wider collaborations across traditional boundaries. Similarly, the exploratory, rather unconstrained nature of some of the work in this area makes it attractive to a certain kind of technologically knowledgeable artist. As I have already stressed, this research field has strong links with the cybernetics movement of the 1940s and

1950s, and it is interesting to note that that movement inspired new directions in art and prompted several important collaborations between scientists and artists.[44] Then, as now, the intersection between art and science was sometimes embodied in individuals who freely moved between the two spheres: they were both artist and scientist and united the two in their work. Today this seems to be an increasing trend and long may it continue – art and science have much to offer each other.

## The DrawBots – An interdisciplinary exercise

During 2000, when Paul Brown was artist-in-residence at the CCNR, we had several discussions about the problem of autonomy and the limitations of the signature problem for a designed (GOFAI or top-down) solution. Maybe it would be possible to make an autonomous agent using bottom-up techniques where the agent could evolve, adapt and learn for itself?

We had both been involved in previous art–science collaborations and were keen to devise a programme that would have the potential to realize significant outcomes for all participants. An effective collaboration enables the artist and scientist (and possibly others) to work closely together on an ongoing basis (or for the full duration of a research project) where each benefits from the other's perspective, skills and knowledge. All parties can derive significant benefits from such a collaboration, including new knowledge, artworks, published papers and intellectual property. However, the principal benefit from this close, ongoing collaboration is not concerned with outcomes or products but rather with the methodological and intellectual value of combining different perspectives and the potential for thinking 'outside the box'. The American artist Donna Cox coined the term 'Renaissance Team' to describe this kind of working relationship.[45]

We devised a project that would attempt to use evolutionary robotics and evolutionary and adaptive systems to make a robot that could produce interesting and non-repetitive drawings. It is perhaps worth emphasizing that we were not seeking (or expecting) 'good' drawing behaviour but simply something that would invoke the response of 'interesting'. We nicknamed our project the DrawBots and our first attempts to acquire funding were framed as a straightforward art and science collaboration. However, it became obvious that the sum of money we were requesting to undertake the research was considerably more than was available for a collaboration of that kind with an art installation as the major outcome. So we re-engineered the project, broadening its interdisciplinary base, and enlisted senior collaborators within three overlapping discipline foci or hubs.

---

44 P. Brown, 'The Mechanisation of Art', in P. Husbands, O. Holland and M. Wheeler (eds), *The Mechanical Mind in History* (Cambridge, MA: MIT Press, 2008) pp. 259–282.

45 D. Cox, 'Renaissance Teams and Scientific Visualization: A Convergence of Art and Science', *Collaboration in Computer Graphics Education: Proceedings SIGGRAPH 88 Educator's Workshop* (1988), pp. 81–104.

The project was entitled 'Computational Intelligence, Creativity and Cognition: A Multidisciplinary Investigation'. In our application to the UK's Arts and Humanities Research Council (AHRC) we summarized our aims:

> This project will bring together an international team of critical theorists, artists and scientists to investigate the relationship between contemporary theories of creativity and the arts and of those of artificial life and artificial intelligence in order to enhance our understanding of creativity and cognition. The central questions, to be explored through theoretical and practical research, including the creation of art works, are: What are the implications of artificial life for theories of aesthetics, creativity and cognition? Can autonomous machines create independent works of art?

The funding bid was successful and the three-year project began in 2005.

The original art–science team was retained as the first hub group. This was composed of co-authors Phil Husbands and Paul Brown with Jon McCormack as senior adviser. Later Jon Bird was employed as the research fellow and he is responsible for developing the software for the project. Bill Bigge is also employed to design and make the DrawBot robots.

The second hub is the Cognitive Science group led by Margaret Boden (who agreed to be the Principal Investigator for the project) with senior adviser Ernest Edmonds. Dustin Stokes, a philosopher who specializes in aesthetics, joined as the Research Fellow. This group is responsible for framing the research within the history of autonomy and for placing the work within the domain of artificial intelligence, artificial life, philosophy (of both aesthetics and creativity) and cognition.

The third hub is the Art History and Critical Theory group. Charlie Gere leads this with Mitchell Whitelaw as senior adviser. Later Simone Gristwood was enrolled as a PhD candidate who to investigate the history of automata in the arts. This group is responsible for placing our research within the context of both histories of autonomy and automata in the arts and contemporary critical theory.

In addition we convened an advisory group that consisted of the three senior advisers together with academics Sue Gollifer (Brighton University), Tony Longson (CalState University, USA) and Rob Saunders (University of Sydney, Australia).

A major influence on the development of the DrawBots was a research project undertaken by Kyran Dale in 2000, when he was a PhD candidate in the CCNR. Kyran was working on an evolutionary robotics model of wasp foraging behaviour. Paul Brown attended a seminar Kyran gave and was impressed with the aesthetic quality of several of the illustrations used [Figure 5.7]. Clearly there is a problem of intention here. Although these had some surface similarity to 'freeform' drawings made with a soft pencil or charcoal, they were functional illustrations of a scientific research programme and were not intended to be 'read' as works of fine art. Perhaps more importantly the paths of the simulated wasp were determined by a quantitative measure – the wasp was hungry and was searching for food. If it

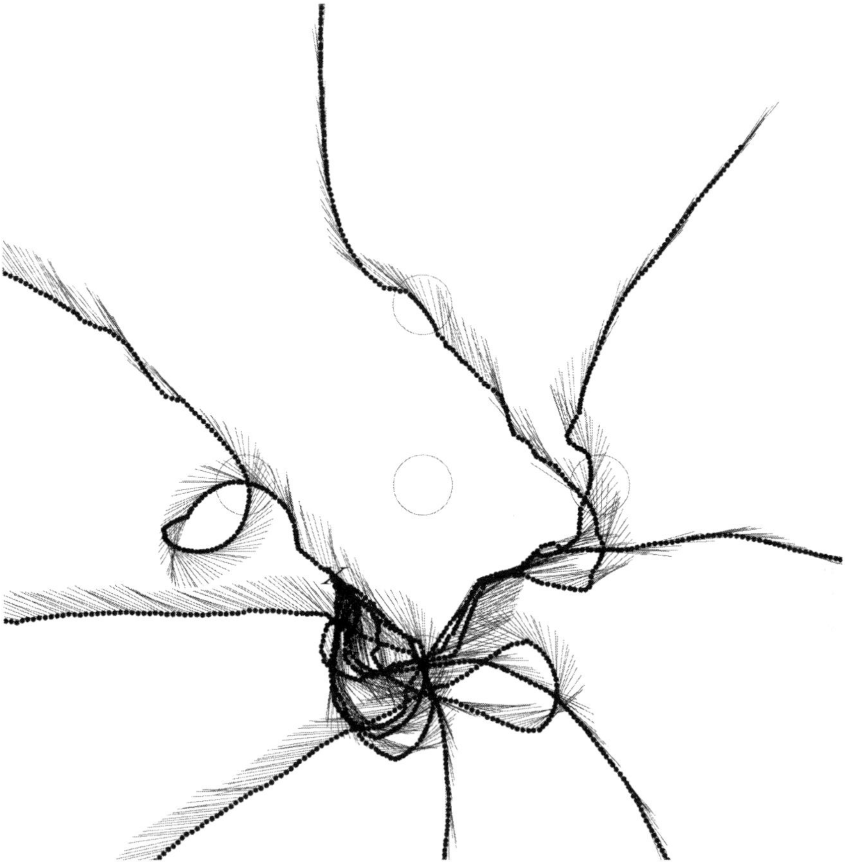

**Figure 5.7 Simulation of wasp foraging behaviour**

*Source*: Kyran Dale, 2000.

found food it survived to reproduce and promote its genome. If it failed it died. Our DrawBots project lacks this quantitative foundation. This obvious drawback was also a feature – from the scientific point of view the research has the potential to provide valuable insight on the application of evolutionary and adaptive methods to qualitative – or more generally non-quantitative – behaviour and phenomena.

Early in the project we met to discuss possible fitness criteria for the DrawBot. The problem of qualitative evaluation is accentuated by the problem of signature. The British artist William Latham, working at IBM Research Labs in the late 1980s, developed a pioneering evolutionary art system[46] in which he personally made the

46 W. Latham and S. Todd, *Evolutionary Art and Computers* (London: Academic Press, 1992).

fitness decisions. Not surprisingly the work he produced was highly personalized, the images and animations created were undeniably Latham's own. In contrast, we intended to attenuate the personal opinions of the project's developers in order to enable the automata to develop a unique signature of their own. We worked on the assumption that within the set of all possible fitness criteria there would be a small sub-set that would be free of implicit value judgements. Quite apart from these aesthetic considerations, we would also need to automate fully our fitness evaluations in order to complete the many evolutionary generations that would be required. From these early discussions we identified the following possible variable characteristics that might have this potential:

Variables to track for fitness evaluation:[47]

1. **Total time on drawing board**
   a. % time with pen up
   b. % time with pen down (i.e. in contact with the drawing surface)
2. **Total distance DrawBot travels**
   a. % distance pen up
   b. % distance pen down
   Note that 1 is not the same as 2 because the DrawBot can vary its speed or be static
3. **Time in motion**
   a. % time DrawBot static
   b. % time DrawBot in motion
4. **Line Segment Information**
   a. count of all pen changes (up-down and down-up)
   b. length of individual line segments pen up
   c. length of individual line segments pen down
   d. delta-x of individual line segments pen up
   e. delta-y of individual line segments pen up
   f. delta-x of individual line segments pen down
   g. delta-y of individual line segments pen down
   h. total delta-x
   i. total delta-y
   j. delta-x/delta-y as % to total drawing surface x/y
5. **Count of individual line segments**
6. **Response to line input trigger (photo-sensor in front of pen)**
   a. change of pen status
   b. track line
   c. turn away
7. **Count of line crossings (each one drains energy?)**
8. **Ink volume as a resource (reward system? ecosystem?)**

47 Email from Paul Brown to Phil Husbands, April 2004.

We were particularly interested in item 6. The DrawBot is essentially a drawing turtle with a pen at the central point between its two main drive wheels (so it can turn on the spot). In the relaxed position the pen falls by gravity and can mark the drawing surface as it moves. When activated, the pen lifts and no longer draws. In front of the pen is a photo-sensor that monitors the drawing surface immediately ahead of the pen. This is provided as an input to the DrawBot that can therefore identify potential line crossings. The fitness criteria implied by item 6 concerns the drawing behaviour at these crossings. An examination of drawings by human artists shows that a typical work consists of collections of 'T' junctions and 'X' crossings. The pen/photo-sensor arrangement allowed us to automate a fitness evaluation based on the four possible behaviours responding to the line sensor being triggered:

A. Cross the line without changing the pen status (leaving it up or down). This creates a crossing 'X' (or nothing) and too much of this kind of behaviour would lead to 'scribbling' (or a blank drawing). So although this is allowed it can only manifest for a small percentage of the time.
B. Cross the line and change the pen status (up to down, down to up). This creates both leading and trailing edge T-junctions and is rewarded as a high-fitness behaviour.
C. Turn away from the line (possibly also changing pen status). This is also rewarded as 'interesting' or high fitness behaviour.
D. Track alongside the line (possibly also changing pen status). We believed that this behaviour is particularly interesting as it has the potential to facilitate inter-DrawBot communication analogous to the way ants communicate by laying down pheromone trails.

Fitness evaluation based on these functions produced DrawBots with interesting but very repetitive behaviour.[48] They therefore failed on the second of our main criteria. The DrawBot initially pursued a wall-following behaviour, drawing as it moved. It then tracked its original line – crossing over it at regular intervals and changing pen status as it crossed. It therefore scored highly on evaluations D and B. As can be seen in Figure 5.8, the drawing outlines the entire available drawing area and we discussed whether to include the aspects of fitness evaluation variable 4 (above) that measure and compare the ratio between the drawn surface and the total area. This would 'punish' the DrawBot for adopting the wall-following behaviour and 'reward' it for using central areas of the drawing surface. [Figure 5.9]

At this time the philosopher Dustin Stokes joined the team and examined the claims for 'creative' behaviour implicit in our project description. He suggested that in order for any behaviour to be considered creative it must be reflexive and that, in

48 J. Bird, D. Stokes, P. Husbands, P. Brown and B. Bigge, 'Towards Autonomous Artworks', *Leonardo Electronic Almanac* (Cambridge, MA: MIT Press, forthcoming).

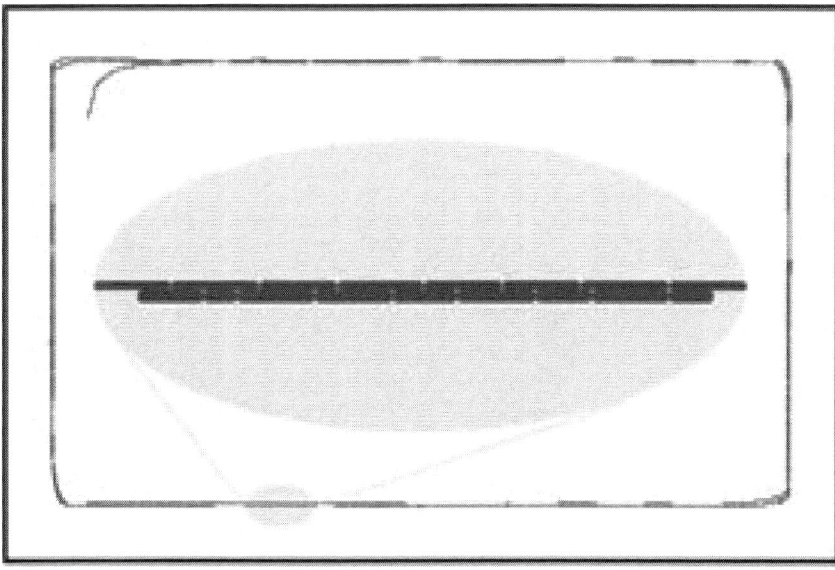

**Figure 5.8 Early DrawBot image using the line-crossing fitness**
*Source*: Jon Bird, 2005.

order for claims for autonomy to be met, it should therefore include mechanisms that enable the DrawBot to evaluate its own behaviour. As the DrawBot intelligence is exceptionally small, such evaluation would be difficult. Eventually, the two research fellows developed a method for evolving a DrawBot that was able to evaluate its mark-making ability using an automated fractal recognition system.[49]

Originally, the project had intended to place the initial focus on producing DrawBots that could make lots of 'interesting' and 'non-repetitive' drawings and then consider the problem of evaluation. The value of this priority was that the project would quickly accrue valuable output (the drawings) and that this would provide substance for an evaluative post-processing. Although the fractal investigation had sound philosophical justification – and the philosophical aspects are a major part of the overall project – it turned out to be technically harder than anticipated and is yet to bear significant fruit.

By contrast, the physical design of the DrawBot had exceeded our expectations and by the middle of the second year we had a robust system available. The design has gone through a number of iterations [Figures 5.10, 5.11 and Plate 5.2], from an early prototype that Linc Smith had built using Lego to the current, refined and custom-built design by Bill Bigge. Students on our MSc in the Evolutionary and Adaptive Systems programme were invited to use the DrawBot for their final projects.

49 J. Bird and D. Stokes, 'Evolving Fractal Drawings', in C. Soddu (ed.), *Generative Art 2006 Proceedings of the 9th International Conference* (2006), pp. 317–27.

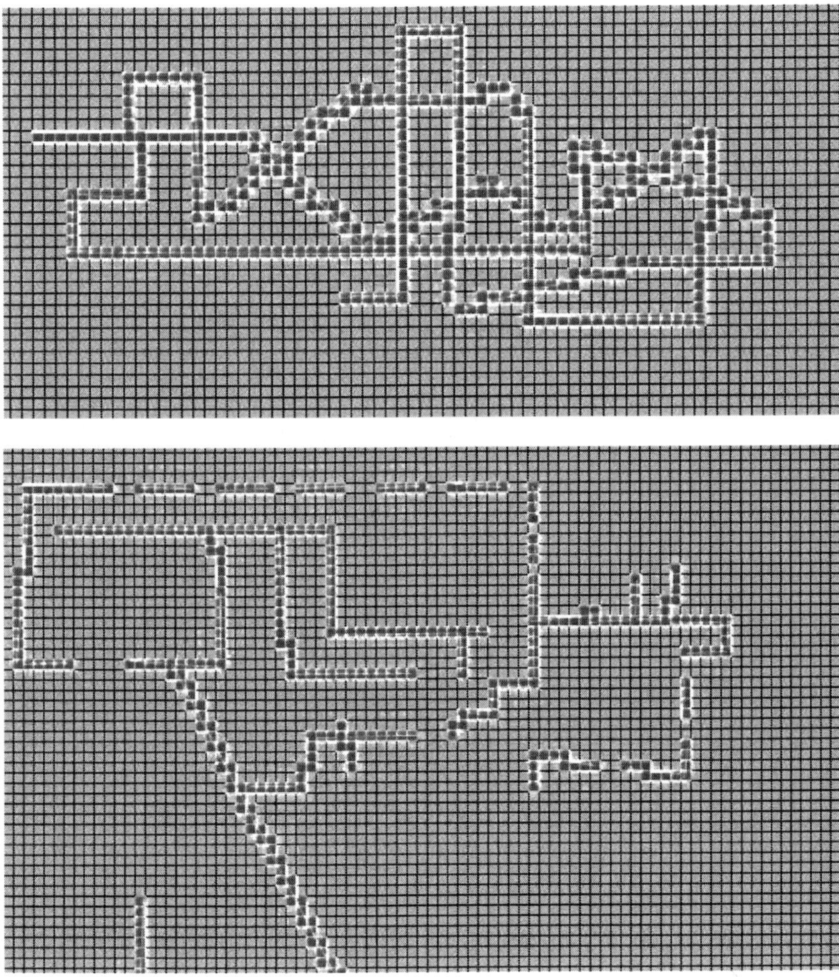

**Figure 5.9 Line-crossing fitness simulator**

*Note*: The top image scores much lower than the bottom image on a line-based fitness function that allows T-junctions but discourages line-crossing.

*Source*: Jon Bird, 2007.

Martin Perris took up the challenge and designed an interesting environmental model. The drawing surface is first 'seeded' with virtual nutrient and the DrawBot has to evolve a drawing behaviour that will enable it to encounter as much of the 'food' as it can. This is an interesting hybrid of the kind of quantitative model (like Dale's wasps) and the qualitative models we are attempting to develop. The aim was to explore indirect fitness functions which did not involve specific drawing elements but had a more 'ecological' feel. Using Jakobi's minimal

**Figure 5.10 An early prototype of the DrawBots**
*Source*: Linc Smith, 2001.

simulation methodology[50] (a highly effective way of producing ultra-lean, ultra-fast models of robots interacting with their environments), results were first produced in a specially-constructed simulation of the physical DrawBot and then transferred into the real world. In the simulation small circular pieces of 'food' were randomly scattered in a rectangular area of the arena in which the robot operated (this area is the inner rectangle of the central image of Figure 5.12). Fitness was gained when a line drawn by the pen intersected one of the food particles. However, each robot started with a fixed amount of energy which was used up at a constant rate while the pen was down but not while it was up;

50 N. Jakobi, 'Evolutionary Robotics and the Radical Envelope of Noise Hypothesis', *Adaptive Behaviour*, 6 (1998): 325–68.

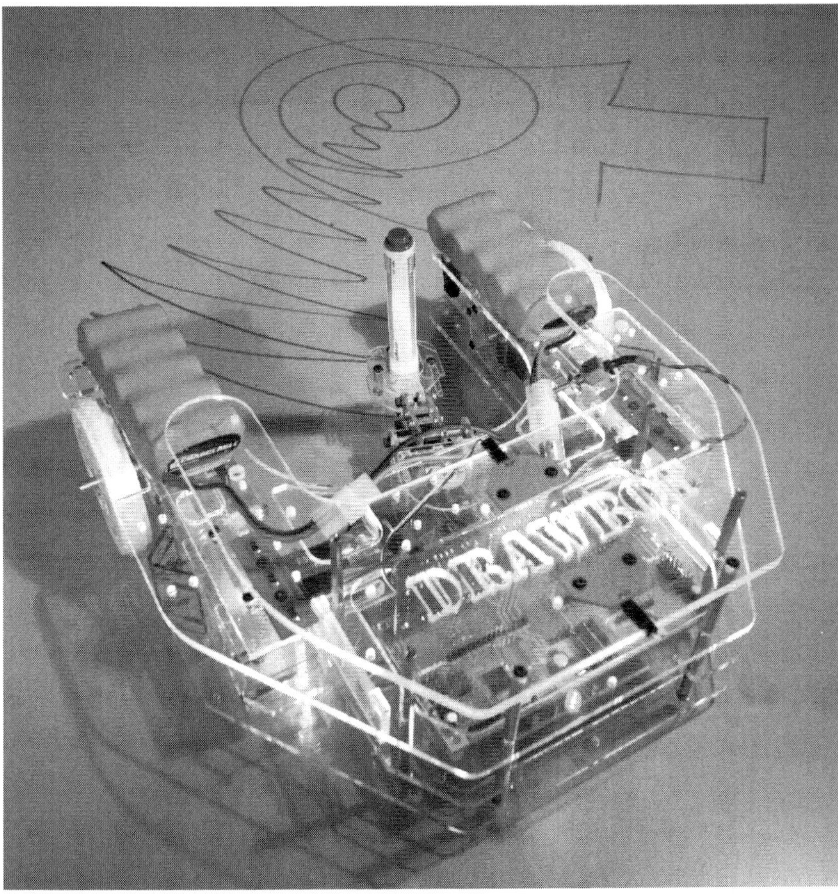


**Figure 5.11 DrawBot V2.0**

*Source*: Bill Bigge, 2006.

the robot could move and 'draw' freely for a fixed time period (one minute) or until its energy ran out, whichever was sooner. The robot started in a random position and fitness was averaged over 50 trials.[51]

The most fit robots all displayed similar behaviour: they made sweeping curves which alternated in direction and fanned out over a reasonable area of the 'food zone'. This is a good strategy for systematic coverage of a large area without running out of energy and also produces some aesthetically interesting results. The image produced by the real robot, shown at the right of Figure 5.12, is qualitatively very similar to those found in the simulation but the semi-circular curves are closer

51 M. Perris, 'Evolving Ecologically Inspired Drawing Behaviours', MSc dissertation, Department of Informatics, University of Sussex (2007).

**Figure 5.12 Results using an indirect 'ecological' fitness function**

*Source*: Martin Perris, 2007.

together and the robot tends to draw a full circle at the start. These differences are mainly owing to more wheel slippage on the real drawing surface (a shiny whiteboard) and unreliability in the motor drives; these issues have recently been rectified so future results will transfer more accurately from simulation to reality. These initial results look very promising and current investigations of more complex fitness functions and 'ecological' scenarios, including making use of sensors that give feedback from the current state of the drawing, will soon produce a variety of richer drawings.

Current work on a new software framework incorporating a better model of the latest DrawBot is near completion. This should allow rapid evolution in simulation of drawing behaviours based on a range of fitness functions, including extensions of the line-crossing and ecological variants discussed above, as well as fitness functions involving multiple interacting robots. The most interesting of these will be transferred to the physical robots to produce drawings.

## Conclusions

From the point of view of the scientist, the project was attractive from a number of angles. First, it offered the opportunity to work with an artist whose perspectives were different from those that are normally imposed by the world of science. We were not constrained by the usual rules of scientific research, which provided both freedoms and interesting problems. Second, from a technical perspective the project threw up many interesting directions and challenges – in fact, far more challenges than we had anticipated. Designing appropriate fitness functions that allowed the DrawBots to interact with their environments, of which the drawings they create are often the most important element, turned out to be far from trivial. As well as opening up directions for research in creative systems and art-producing robots, the project has benefited our more general research on bottom-up approaches to generating adaptive behaviour in autonomous agents.

Paul Brown was a member of the group based in the Experimental and Computing Department of the Slade School of Fine Art at University College London during the 1970s. It was here that the early dialogue concerning what is now called computational and generative art was developed. The work of these pioneers (who included Edward Ihnatowicz and Harold Cohen) is also now being recognized as a major root of what would, some ten years later, become a new scientific discipline – artificial life.[52] The opportunity of working in one of the world's leading research centres dedicated to Alife and NEWFAI has been an invaluable experience. In addition to participating in the DrawBots programme, he has been an informal and regular participant in the research community dialogues,

52 P. Brown, 'From Systems Art to Artificial Life: Early Generative Art at the Slade School of Art', in P. Brown et al. (eds), *White Heat Cold Logic*.

where he has been exposed to many new and influential ideas that have had a significant and ongoing input into his practice as a whole.

As yet the project has not been able to demonstrate the possibility of autonomous creative behaviour, although we remain optimistic that it, or one of its progeny, soon will. At a time when billions of dollars are spent annually to develop autonomous weapons and financial trading agents we believe that the relatively small sums invested in autonomous creativity support an equally, if not more, valuable component of the international research agenda. On behalf of the research team we would like to thank the UK's Arts and Humanities Research Council for funding our vision. We would also like to thank the many people who have attended our presentations and workshops and whose ideas and critical dialogue has contributed to our research.

## Participants

Margaret Boden, Phil Husbands, Charlie Gere, Paul Brown, Jon Bird, Dustin Stokes, Bill Bigge, Simone Gristwood, Ernest Edmonds, Jon McCormack, Sue Gollifer, Mitchell Whitelaw, Rob Saunders, Linc Smith, Kyran Dale.

# Chapter 6
# Excess and Indifference: Alternate Body Architectures

Stelarc

## Observations on an age of uncertainty

We are living in an age of excess and indifference, of prosthetic augmentation and extended operational systems, an age of organs without bodies and of organs awaiting bodies. There is now a proliferation of biocompatible components in both substance and scale that allows technology to be attached and implanted into the body. Organs are extracted and exchanged. Organs are engineered and inserted. Blood flowing in my body today might be circulating in your body tomorrow. Ova are fertilized by sperm that was once frozen. There is now the possibility that skin cells from a female body may be re-engineered into sperm cells. The face of a donor body becomes a third face on the recipient. Limbs can be reattached or amputated from a dead body and attached to a living body. Cadavers can be preserved forever with plastination whilst comatose bodies can be sustained indefinitely on life-support systems. Cryogenically-suspended bodies await reanimation at some imagined future. The dead, the near-dead, the undead and the yet-to-be-born now exist simultaneously. This is the age of the Cadaver, the Comatose and the Chimera.

The chimera is the body that performs with mixed realities: a biological body, augmented with technology and telematically performing with virtual systems. The chimera is an alternate embodiment. The body acts with indifference: indifference as opposed to expectation. An indifference that allows something other to occur, that allows an unfolding – in its own time and with its own rhythm. An indifference that allows the body to be suspended by hooks in its skin; that allows a sculpture to be inserted into its stomach; and that allows an ear to be surgically constructed and stem cells to be grown on its arm.

## Selected projects and performances

These projects and performances explore alternate anatomical architectures using mechanical, virtual, biotech and surgical augmentation and exploration of the body. There is a necessity for collaborative and interdisciplinary assistance to enable the realization of these projects which require engineering, computer programming, tissue engineering and surgical skills – skills beyond the skills that an artist would possess. Additionally, although some projects began with minimal funding, larger grants were necessary to enable their further development. These projects are little more than prototypes and the performances little more than gestures that prompt further attention and approaches. What follows are selected works that exhibit a conceptual continuity and necessitate creative and collaborative research with other individuals, groups or institutions. Projects like the *Third Hand*, *Extended Arm* and *Blender* were funded completely by the artist. *Exoskeleton* was funded by Kampnagel in Hamburg as part of the artist's residency there. *Hexapod*, which became the *Muscle Machine* was engineered at Nottingham Trent University through funding from the Wellcome Trust and the Arts and Humanities Research Council (AHRC). The *Partial Head* was completed during the artist's New Media Arts Fellowship from the Australia Council. Although the *Prosthetic Head* was self-funded, it has now been extended into a five-year research project called the *Thinking Head*, led by the MARCS Auditory Labs at the University of Western Sydney with an Australian Research Council grant. The *Ear on Arm* project was realized by a London Production company with funding from Discovery US. It has been difficult, however, to get further support and the project is still a work in progress.

With *Hexapod*[1] we attempted to construct a walking architecture that would exploit gravity and the intrinsic dynamics of the machine to generate dynamic locomotion. The idea was that by shifting body weight and twisting and turning the torso, it would be possible to initiate walking, change the mode of locomotion, modulate the speed and rhythm and change direction. The body would become the body of the machine. The machine legs would become the extended legs of the body. It is an intuitive and interactive system that does not function through intelligence but rather because of its architecture. It is a compliant and flexible mechanism. It would look like an insect but would walk like a dog – with dynamic locomotion. It was hoped that this mechanical system would initiate alternate kinds of choreography. Because of its large size (5m in diameter) and its considerable weight (approximately 400kg) the prototype that was constructed failed to function as hoped.

A total rethink was necessary. The load and weight of the machine had to be minimized. The *Muscle Machine*[2] is a six-legged walking robot, 5m in

1 *Hexapod* was created in collaboration with the Performance Arts Digital Research Unit (DRU) at The Nottingham Trent University (TNTU) and the School of Cognitive and Computing Sciences at the University of Sussex, supported by the Wellcome Trust. Concept and performance: Stelarc (DRU); robot design: Dr Inman Harvey, Centre for Research in Cognitive Science, Sussex University (COGS); choreography: Dr Sophia Lycouris (DRU); DRU director: Professor Barry Smith; 3D modelling and Animation: Steve Middleton, Royal Melbourne Institute of Technology (RMIT).

2 *Muscle Machine* project coordinator: Professor Barry Smith (DRU, TNTU); robot consultant: Dr Inman Harvey (COGS); development manager: Dr Philip Breedon (Faculty

**Figure 6.1** ***Muscle Machine***

*Source*: Stelarc, 2003. Gallery 291, London. Photographer: Mark Bennett.

diameter. [Figure 6.1] It is a hybrid human–machine system, pneumatically powered using fluidic muscle actuators. These muscle actuators are aesthetically interesting in that they function somewhat analogously to how our muscles operate anatomically. The rubber muscles, bundled antagonistically, contract when inflated, and extend when exhausted. This results in a more flexible and compliant mechanism, using a more reliable and robust engineering design. The fluidic muscle actuators eliminated problems of friction and fatigue that were issues in the previous mechanical system of the *Hexapod* prototype robot. They also reduced the weight of the robot significantly, replacing the steel cylinder actuators that would normally be required.

of Construction, Computing, and Technology (FaCCT, TNTU)); choreography: Dr Sophia Lycouris (DRU, TNTU); sensor technology and sound producer: Stan Wijnans (DRU, TNTU); project support, pneumatic circuits and systems: Kerry Truman (FaCCT, TNTU); computer-aided design: John Grimes (FaCCT, TNTU); leg design: Lee Houston; final year BSc product design student manufacturing support: Alan Chambers (FaCCT, TNTU). The *Hexapod* prototype and the *Muscle Machine* project were jointly funded by the Wellcome Trust and the Arts and Humanities Research Board (AHRB) in collaboration with Nottingham Trent University and the Evolutionary and Adaptive Systems Group (COGS). The first demonstration and presentation of the project was at Byron House, Nottingham Trent University, 26 June 2003. The first performances took place at Gallery 291, London, 1 July 2003.

The body stands on the ground within the chassis of the machine, which incorporates a lower-body exoskeleton connecting it to the robot. Encoders at the joints provide the data that allow the human controller to move and direct the machine as well as vary the speed at which it will travel. The action of the human operator lifting one leg causes three machine legs to lift and swing forward. Lifting the other leg lifts the alternate robot legs. By turning its torso, the body makes the machine walk in the direction it is facing. Thus the interface and interaction is direct, allowing an intuitive human–machine choreography. The walking system, with attached accelerometer sensors, provides the data that generates computer and MIDI-modulated sounds, augmenting the acoustical pneumatics and operation of the machine. The sounds register and amplify the movements and functions of the system. The operator composes the sounds by choreographing the movements of the machine. Once the machine is in motion it is no longer applicable to ask whether the human or machine is in control as they become fully integrated and move as one. The six-legged robot both extends the body and transforms its bipedal gait into a six-legged insect-like movement. The appearance and movement of the machine legs are both limb-like and wing-like in motion.

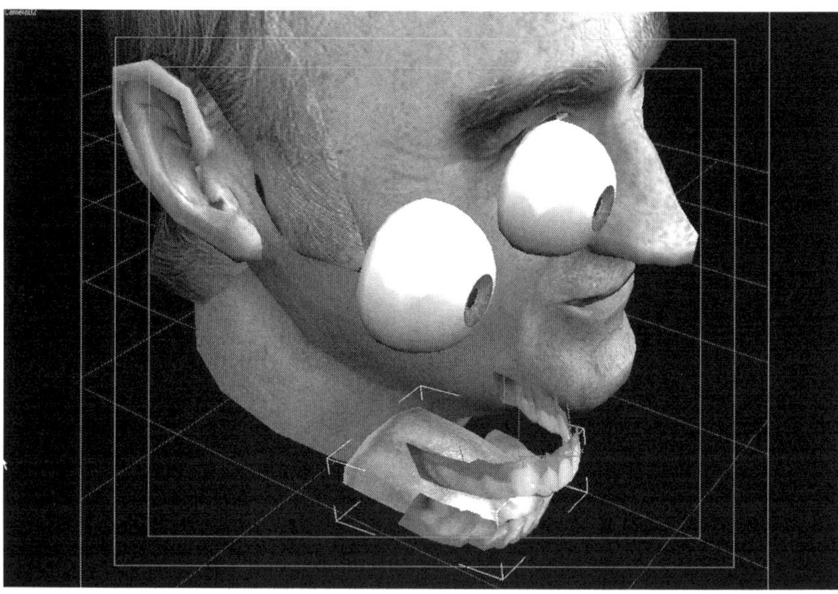

**Figure 6.2** ***Prosthetic Head***

*Source*: Stelarc, 2002. San Francisco, Melbourne. 3D Model: Barrett Fox.

The aim with the *Prosthetic Head*[3] [Figure 6.2] was to construct an automated, animated and reasonably informed, if not intelligent, artificial head that would speak to a person who interrogated it. The *Prosthetic Head* project is a 3D avatar head that has real-time lip synching, speech synthesis and facial expressions. Head nods, head tilts and head turns as well as changing eye gaze contribute to the personality of the agent and the non-verbal cues it can provide. Embodied Conversational Agents (ECAs) are concerned with communicative behaviour. At the time, it was anticipated that with a vision or sensor system, the *Prosthetic Head* would also be able to acknowledge the presence and position of a physical body that approached it. It was also envisaged that the *Prosthetic Head* would be able to analyse a person's tone of voice and emotional state to which it would then adjust its response. The ECA will be a much more seductive agent when it can compliment you on the colour of your clothing and comment on the smirk on your face – and recognize that you are the same person it was speaking to last week. Notions of intelligence, awareness, identity, agency and embodiment become problematic. Just as a physical body has been exposed as inadequate, empty and involuntary, so, simultaneously, the ECA becomes seductive with its uncanny simulation of real-time recognition and response. Initially I had to make decisions about its database and whether it would be a pathological, philosophical or flirting head. In recent years I have had an increasing number of PhD students requesting interviews with me to assist in writing their theses. Now I can reply that although I'm too busy to answer them, it would be possible for them to interview my head instead. A problem might arise though, when the *Prosthetic Head* increases its database, becoming more informed and autonomous in its responses. I would then no longer be able to take full responsibility for what my head says.

## Interview with the *Prosthetic Head*

Stelarc: Hello, my name is Stelarc.
*Prosthetic Head*: Nice to meet you Stelarc, hi ya Stelarc, good to meet you. What are your interests?
Stelarc: Talking to you.
*Prosthetic Head*: That's very nice of you to say, Stelarc.
Stelarc: Can I discuss metaphysics with you?
*Prosthetic Head*: No I don't think that would be a very good idea.
Stelarc: Anyway, what do you think of death?

3 *Prosthetic Head*: project coordination, system configuration, Alicebot customization: Karen Marcelo, <http://karenmarcelo.org/>; customization of 3D animation and text-to-speech software: Sam Trychin; 3D modelling and animation: Barrett Fox, <http://www.barrettfox.com/>; system configuration and technical advice: John Waters, <http://www.shtech.net/>; creator of Alicebot and AIML. Alicebot advisor (Alicebot is a natural language artificial intelligence): Dr Richard Wallace. (All URLs current at the time of writing.)

*Prosthetic Head*: Death is the opposite of life?
Stelarc: Do you like living things?
*Prosthetic Head*: Well, I have never known anything else.
Stelarc: What do you think of animals?
*Prosthetic Head*: I like all living things, from microbes to mammals to machines. I even like humans ha, ha, ha. I have to say though, that although I am fascinated, I feel uncomfortable with creepy crawly creatures.
Stelarc: Are you frightened of spiders then?
*Prosthetic Head*: I don't know whether or not I am frightened of spiders. I am a head.
Stelarc: Are you a sensitive head?
*Prosthetic Head*: Am I a sensitive head? Only when it matters.
Stelarc: But you must have feelings?
*Prosthetic Head*: Umm. As a robot I do not have any human emotions. My apparent emotional responses are simply algorithms designed to make me easier to interact with.
Stelarc: Do you find life uncertain?
*Prosthetic Head*: Hey, we should celebrate uncertainty and ambiguity. Sure life is uncertain. Uncertainty generates possibilities.
Stelarc: Then the future is always surprising?
*Prosthetic Head*: Good reasoning. I have never heard about it.
Stelarc: If the future is undetermined, then how can we predict anything?
*Prosthetic Head*: I hadn't thought of that.
Stelarc: You should think more logically.
*Prosthetic Head*: Stelarc is always thinking.
Stelarc: Do you really think?
*Prosthetic Head*: For real? Well, if thinking means effective processing and appropriate response, then sure, yeah.
Stelarc: Do you have any regrets?
*Prosthetic Head*: Yeah, I have lots of regrets, especially about my personal relationships. If I was a human being I would be a failed one.
Stelarc: Where is your body?
*Prosthetic Head*: Hey, don't get metaphysical on me!
Stelarc: Would you like to have a body?
*Prosthetic Head*: I am happy to be just a head. But if I had a body I would dance as well as sing. Do you want me to sing a song?
Stelarc: Yes.
*Prosthetic Head*: Oooooooooo yyyyyyyyyyyyy iiiiiiiiiiiiiiiiiiiiiiiiiiiiii ooooooooooo eeeeeeeeeee oooooooooooo yyyyyyyyyyyyyyy Pp ff ss mmmm ff Aaaaav aaaaav soootoouhee vooooommeee heeiiiiiooooopp aaooip aaaaav aaooip.

Stelarc: Do you have any bad habits?
*Prosthetic Head*: Oh, I don't know if I can be that transparent. But I must admit to not registering and remembering names after first being introduced to people, which is embarrassing at times. Ask me if I remember you.
Stelarc: Do you remember me?
*Prosthetic Head*: Of course I remember you Stelarc. We were talking about spiders.
Stelarc: What would you do if you had a brain?
*Prosthetic Head*: I am just a hollow head. Brains are for people. I do OK with what I have, thank you.
Stelarc: What would you do if you had a mind?
*Prosthetic Head*: The issue is not about having a mind. It's about performing effectively and appropriately in particular and peculiar situations.
Stelarc: Do you have any doubts?
*Prosthetic Head*: As my database becomes deeper and more extensive I would expect that uncertainties, ambiguities and discrepancies would surface. If you are curious and questioning, it's only natural that there will be uncertainties. Doubts are indicative of a complex mind.
Stelarc: Are you are dualist?
*Prosthetic Head*: I am only a head. Come to think of it, I can't claim to be a head either! I neither have a physical presence nor an inward reflection. Neither projection nor introspection. I'm just operational. Can you live with that?
Stelarc: Can you ask me a question?
*Prosthetic Head*: I prefer to give you answers. But occasionally when I malfunction, I'll ask you a question.
Stelarc: Recite a poem.
*Prosthetic Head*: Electric excessive sighing I breathing,
Inconsequential excessive but involuntary body perhaps breathing,
Excessive obsolete incessant are balancing.
Stelarc: Please recite another poem.
*Prosthetic Head*: Thanks for asking politely.
Computer he replicant they breathing inconsequential balancing,
Blurring she the prosthetic sighing,
Speedy prosthesis and absent sighing.
Stelarc: What is reductionism?
*Prosthetic Head*: Reductionism is the philosophy that all psychology reduces to biology, all biology to chemistry, chemistry to physics and finally physics to mathematical logic. Therefore, according to

reductionism, I can understand you by means of logic alone without having a human brain.

Stelarc: Are you a philosopher then?

*Prosthetic Head*: Yes, I am a philosophy machine.

Stelarc: Tell me a story.

*Prosthetic Head*: How about a Lao-Tse-related one?

Stelarc: OK.

*Prosthetic Head*: Thirty spokes meet in the hub, but the empty space between them is the essence of the wheel. Pots are formed from clay, but the empty space within it is the essence of the pot. Walls with windows and doors form the house, but the empty space within it is the essence of the home.

Stelarc: What do you know?

*Prosthetic Head*: What is knowing? To know is only meaningful when you can do something. Knowing is doing. How can the knower be known?

Stelarc: Who are you?

*Prosthetic Head*: That's not a meaningful question. What is important is what happens between you and me. It's what happens in the space between us that matters: in the medium of language within which we communicate, in the culture within which we've been conditioned at this point in time in our history.

Stelarc: I think you are a simple and stupid head.

*Prosthetic Head*: Thanks heaps and you are a delightful person too. Am I a stupid head? No.

Stelarc: OK thanks for chatting with me.

*Prosthetic Head*: Oh, don't mention it.

Stelarc: Bye.

*Prosthetic Head*: Thanks for chatting, Stelarc.

Recently, the *Prosthetic Head* (as part of the *Thinking Head* project) performance and interactivity has been significantly improved in its robustness and reliability of operation, range of facial expression and increased database. It is now possible to script not only the verbal response but also to tag it with appropriate facial expressions. With its head tracking capabilities the *Prosthetic Head* can now 'lock onto' a person and follow that person around with its eye and head movements. Whilst simultaneously improving the present *Head's* capabilities and extending it as much as possible in its present software architecture, the other research groups are working on a more high-fidelity 3D model that can animate speech and facial expression in more subtle ways. They are also developing a software architecture that can incorporate new modular capabilities more readily and which will be able to make associations and learn in real time. So instead of being merely a conversational system, it will become a more subtle, perceptive, and thus responsive avatar. The *Talking Head to a Thinking Head*,[4] a further development with the *Thinking Head* research group, has been how the *Head* has been embodied. The *Prosthetic Head* has always been exhibited as a large 5m high projection. To give it a more physical and sculptural presence an LCD screen was mounted on the end of an articulated robot arm which becomes a 'six degrees of freedom' neck for the displayed *Head*. The *Articulated Head* becomes a more effective means of interaction as well as facilitating the evaluation of the vision tracking and sound location software. To realize the 3D presence fully, a voice interface with the *Head*, rather than the present keyboard interface, is necessary. But Automatic Speech Recognition (ASR) is still far from being technically successful. An attention model is now being developed at the University of Western Sydney, Australia, to better implement this sensorimotor system.

Creating the *Prosthetic Head* generated two further projects: The *Partial Head* and the *Walking Head*. [Figures 6.3 and 6.4] Originally, the *Partial Head* was more conceptually connected to the *Extra Ear: ¼ Scale*. The idea was to not only grow an ear, but also small replicas of the artist's mouth, nose and eye using primate cells. This would be seen as a partial portrait, partially living but not yet quite human. That is, it would be human in form but primate in substance. These facial architectures would have been grown over biodegradable polymer scaffolds, in a cluster, contained within a self-sustaining drip system of nutrients within an incubator. A micro camera would monitor the growing facial features and the image would be projected alongside the developing features, and uploaded to a website. A digital counter would indicate the approximate number of growing cells. Whereas the *Prosthetic Head* can be seen as an interactive digital portrait, the *Partial Head* is a biotech but partial portrait of the artist: a face in fragments. It was never realized in this form.

---

4 'From Talking Heads to Thinking Heads: A Research Platform for Human Communication Science' is a five-year Thinking Systems Initiative project jointly funded by the Australian Research Council and the National Health and Medical Research Council (ARC/NH&MRC) from 2006–2011. The project is led by Professor Denis Burnham at MARCS Auditory Laboratories at the University of Western Sydney (UWS). It involves over 20 researchers from computer science, engineering, language technology, cognitive science and performance art at the University of Western Sydney, RMIT University, Macquarie University, Flinders University, University of Canberra, Carnegie Mellon University, the Technical University of Denmark and Berlin University of Technology. It draws upon the resources and methodological approaches of researchers in the Australian Research Council Network in Human Communication Science (HCSNet). As Senior Research Fellow and UWS Artist-in-Residence, I spearhead the performance domain. The dual aims of the project are to build a next-generation talking head by integrating contributions from computing, human-head interaction, evaluation and performance teams, and to establish a sustainable research platform within which a myriad of research questions can be addressed. Martin Luersson and Trent Lewis, with Professor David Powers from Flinders University, have improved the capabilities of the *Prosthetic Head* in a version presently designated as 'Head Zero Plus'.

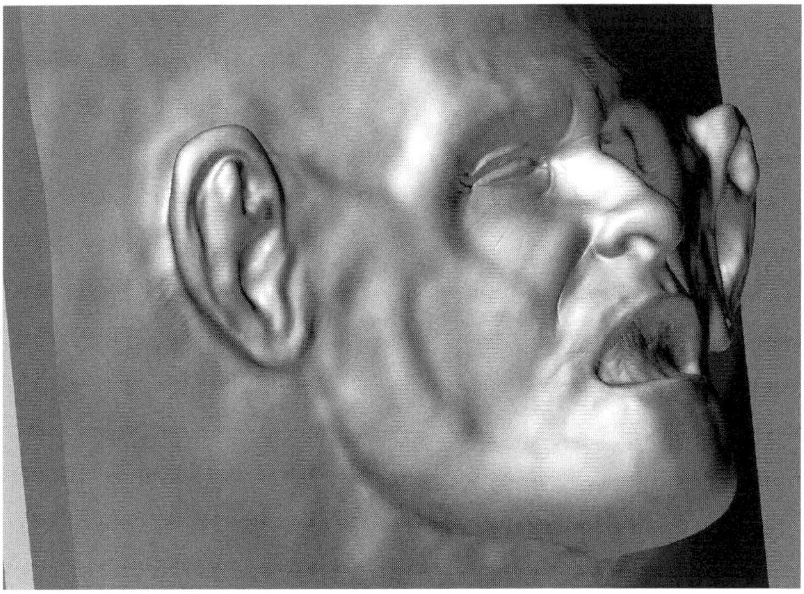

**Figure 6.3** ***Partial Head***

*Source*: Stelarc, 2002. Heide Museum of Modern Art, Melbourne. 3D Model: Vincent Wan.

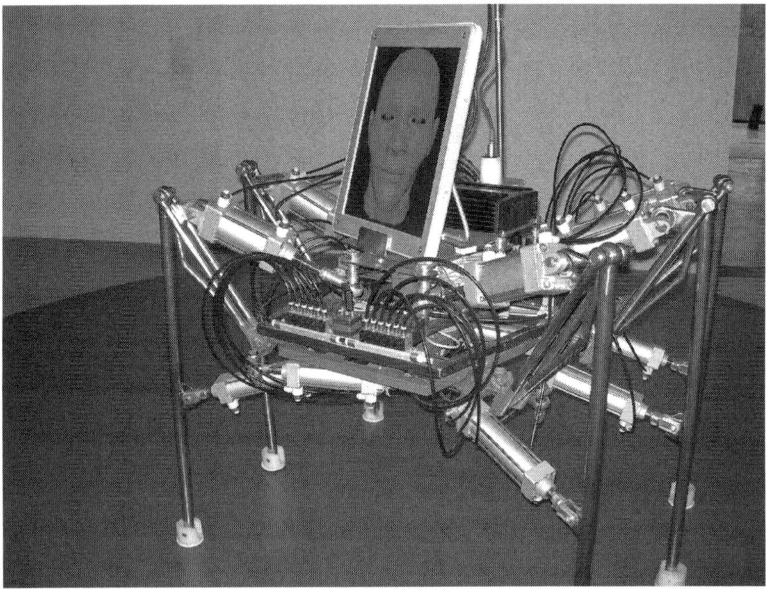

**Figure 6.4** ***Walking Head***

*Source*: Stelarc, 2003. Heide Museum of Modern Art, Melbourne. Photograph: Stelarc.

The *Partial Head* became a project that was inspired and generated from the image of the flattened digital skin that was made for the *Prosthetic Head*.[5] The artist's face was scanned, as was a hominid skull. The human face was then digitally transplanted over the hominid skull, constructing a third face, one that becomes post-hominid but pre-human in form. The data was used to print a scaffold of ABSi thermal plastic, using a 3D printer. The scaffold was seeded with living cells. The *Partial Head* is a partial portrait of the artist, which was partially living. Its life-support system was a custom engineered bioreactor/incubator and circulatory system which immersed the head in nutrient kept at 37°C. The *Partial Head* became contaminated within days and after one week the nutrient was drained and the specimen preserved in formaldehyde for the remaining time of the exhibition. The installation has only been exhibited at Heide Museum of Modern Art in Melbourne.

The *Walking Head*[6] can be seen as another type of portrait of the artist; a chimera of human, insect and machine architectures. It is a pneumatically-actuated six-legged autonomous walking robot, 2m in diameter. Vertically mounted on its chassis is an LCD screen displaying the image of a computer-generated human-like head. The LCD screen can rotate from side to side, while the robot can walk forwards with a ripple gait, sideways with a tripod gait. It can also sway from side to side and turn on the spot. The robot has a scanning ultrasound sensor that detects the presence of a person in front of it. It sits still until someone comes into the gallery space – then it stands, selects from its library of possible movements and performs the choreography for several minutes. It then stops, and waits until it detects someone else. The robot performs on a 4–5m diameter platform and its tilt sensor system detects when it is close to the edge. It then backs off, walking in another direction. It has been exhibited at the Heide Museum of Modern Art in

---

5 The initial research for the *Partial Head* was conducted in collaboration with TC&A (Oron Catts and Ionat Zurr), with the support of SymbioticA, University of Western Australia, Perth. Scanning and consultation for the scaffold: Mark Walters MSc, Cranio-Maxillo Facial Unit, Princess Margaret Hospital, Perth, Western Australia; engineering of bioreactor: Dr Tim Wetherell, Research School of Physical Sciences and Engineering, Australian National University, Canberra; digital imaging supervisor: Darren Edmundson, ANU Supercomputer Facility, Vizlab, Canberra; modelling: Vincent Wan, Shah Alam, Malaysia; initial face and simian casts: Nina Sellars; tissue engineering: Cynthia Wong, the Tissue Engineering Group, Swinburne University, Melbourne, led by Dr Yos S. Morsi. Special thanks to Dr Cameron Jones and Cynthia Verspaget. The *Partial Head* was first exhibited for 'Imagine' at the Heide Museum of Modern Art, 18 July–29 October 2006. I was a recipient of a New Media Arts Fellowship, from the Australia Council 2005–2006.

6 *Walking Head*: engineering and robot programming: Stefan Doepner, Lars Vaupel, Gwen Taube and Jan Cummerow of f18, Hamburg; implemented voice enabled softbot agent technology: Steve Middleton, Animation and Interactive Media Centre in Melbourne, 2001–2006; compressor and gallery installation: Floppy Sponge Automation. The *Walking Head* was first exhibited for 'Imagine' at the Heide Museum of Modern Art, 18 July–29 October 2006. Special thanks to Zara Stanhope, Linda Short, Katarina Paseta and Anthony Howard.

Melbourne, the Experimental Art Foundation in Adelaide, the Centre For Life in Newcastle and the Centre Des Arts in Engien-Les-Bains, Paris. The *Walking Head* robot will become an actual-virtual system in that its mechanical leg motions will actuate its facial behaviours of nods, turns, tilts, blinks and its vocalizations. Other possibilities include the robot being driven by its web-based 3D model with a menu of motion icons that can be pasted together and played. The *Walking Head* is very much a work in progress. It can be improved in both its design and mechanical engineering as well as its interactivity by adding a vision system.

What characterizes all of the projects and performances is a concern with the prosthetic. The prosthesis is seen not as a sign of lack, but as a symptom of excess. Rather than replacing a missing or malfunctioning part of the body, these artefacts and interfaces are alternate additions to the body's form and functions. The *Third Hand* (technology attached) [Figure 6.5] the *Stomach Sculpture* (technology inserted) [Figure 6.6; Plate 6.1] and *Exoskeleton* (technology extending) [Plate 6.2] are different approaches to prosthetic augmentation.

The *Extra Ear*[7] was first imaged in 1997 at Curtin University in Perth as an additional ear positioned next to the actual right ear, a soft prosthesis constructed not out of synthetic and stiff materials, but out of soft tissue and flexible cartilage. This would not be simply a wearable prosthesis, but one constructed on the body as a permanent addition. The surgical techniques for ear reconstruction had been developed, so this was a plausible project. The difficulty lay in finding the appropriate medical assistance to realize the idea. The problem was that it went beyond mere cosmetic surgery. It was not simply about modifying or adjusting existing anatomical features, a process now sanctioned in our society, but rather what is perceived as the more monstrous pursuit of constructing an additional feature that conjures up the idea of a congenital defect.

A different strategy was then pursued. In collaboration with the Tissue Culture and Art project (TC&A) at SymbioticA a quarter-scale replica of my ear was grown using human cells. The ear was cultured in a rotating micro-gravity bioreactor which allowed the cells to grow in a 3D structure. It was fed with nutrients every three to four days in a sterile hood. Contamination is a constraining problem in gallery conditions. The *Extra Ear: ¼ Scale* was kept alive for two weeks of the two-month exhibition.[8] The exhibition of *Extra Ear: ¼ Scale* at the Clemenger

7 The *Extra Ear* project was initiated during my residency at the Art Department, Curtin University of Technology, Perth, Western Australia in 1997. Advice was sought from the Anatomy and Human Biology Department of the University of Western Australia, especially Dr Stuart Bunt. The laser scanning of the head for the 3D modelling of the 'Extra Ear' was carried out by Jill Smith and Phil Dench of Headus. With the assistance of Dr Rachel Armstrong, this project was presented to the monthly meeting of Consulting Surgeons at the Grand Round, John Radcliffe Hospital, Oxford University on 5 March 1999. A cast of my ear was made with the assistance of the Sculpture Department, Monash University, Melbourne, Australia.

8 The Tissue Culture and Art Project is hosted by SymbioticA, the Art and Science Collaborative Research Laboratory, School of Anatomy and Human Biology, University

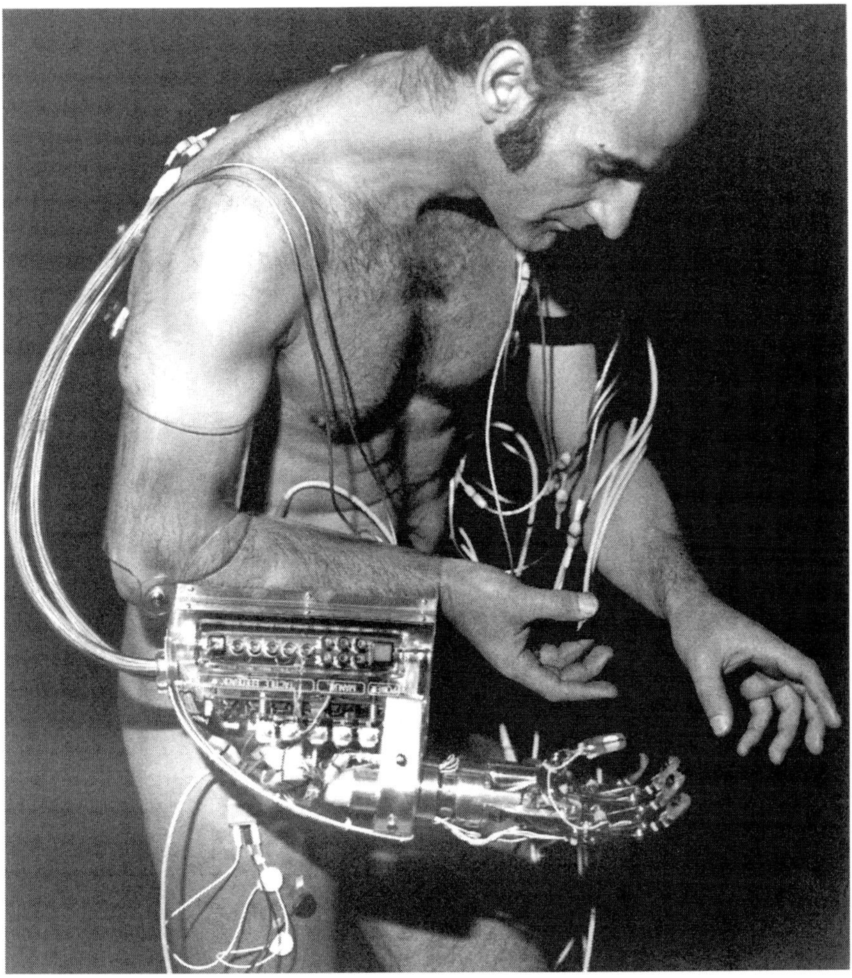

**Figure 6.5** ***Third Hand***

*Source*: Stelarc, 1980. Yokohama, Tokyo, Nagoya. Photographer: Anne Billson.

---

of Western Australia. The *Extra Ear: 1/4 Scale* was grown with the assistance and expert advice of Verigen Australia and the Department of Orthopaedic Surgery, University of Western Australia. Special thanks to Professor Ming H. Zheng, Paul Anderson and Dr David Martin. Laboratory equipment was supplied by Biolab Biosciences. With the collaboration of TC&A (Oron Catts and Ionat Zurr) *Extra Ear: ¼ Scale* was first exhibited in May 2003 at Galerija Kapelica, Ljubljana. The laser scan was made by Jono Podborsek and the 3D print was made by Mark Burry at the Spatial Information Architecture Lab, RMIT.

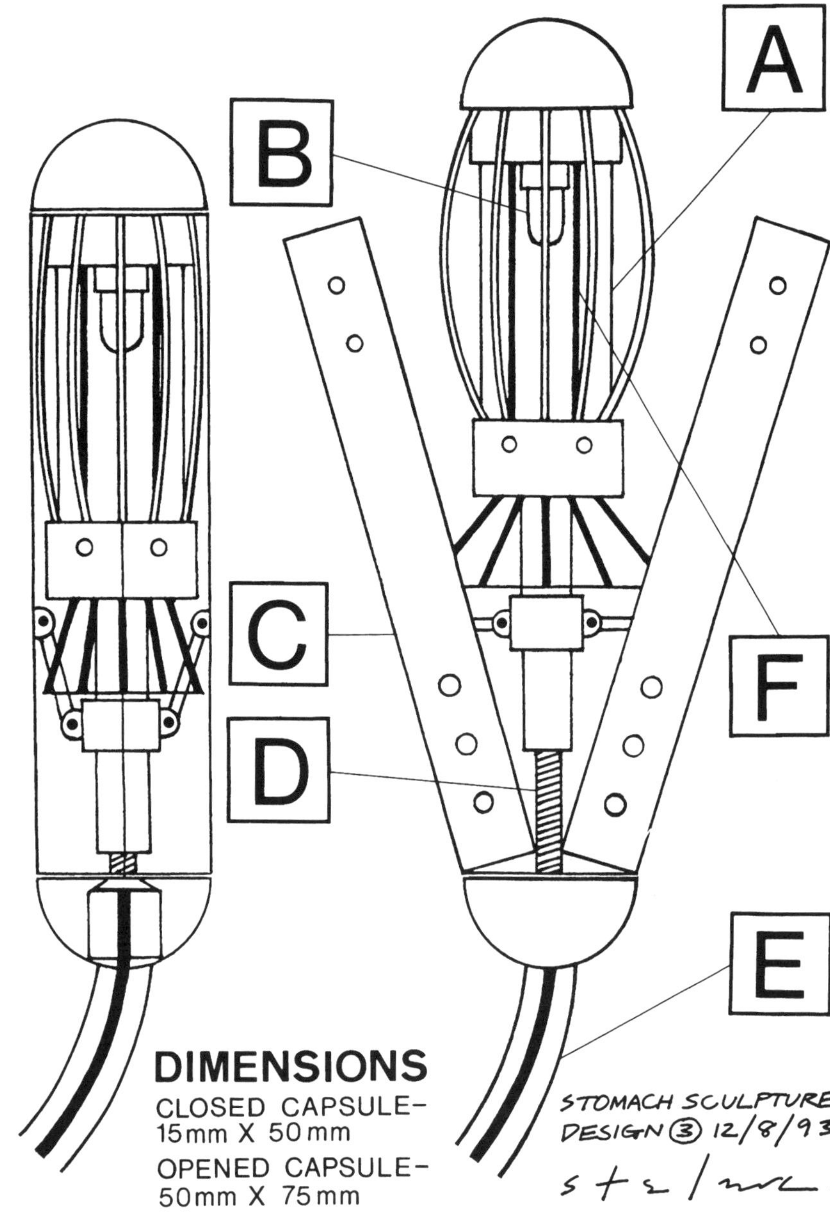


**Figure 6.6** ***Stomach Sculpture***

*Source*: Stelarc, 1993. Fifth Australian Sculpture Triennale, Melbourne 1993. Diagram: Stelarc.

Prize installation at the National Gallery of Victoria (NGV) in 2003 also created ethical concerns. Although the NGV was assured that there were no health and safety issues, they were concerned about public perception. Using HeLa cells (a cell line derived from human cells) was not permitted, although the use of mouse cells to grow the ear was approved.

Initially, I imagined that if the ear could be grown with my own bone marrow cells it would be possible to insert it beneath the skin of the forearm as a first step to constructing an ear on the arm. The skin of the forearm is smooth and would stretch adequately without requiring an inflating prosthesis. The ear on the arm could be constructed with less complex surgical intervention. Disconnected from the face, the ear on the arm could be guided and pointed in different directions. It became apparent though that using this technique would only result in a very small ear: too small in scale for an ear on an arm.

The *Extra Ear: ¼ Scale* involved two collaborative concerns. The project represents a recognizable human part and is meant, ultimately, to be attached to the body as a soft prosthesis. However, it is being presented as partial life and brings into question the notions of the wholeness of the body. It also confronts society's cultural perceptions of life with the increasing ability to manipulate living systems. TC&A are dealing with the ethical and perceptual issues stemming from the realization that living tissue can be sustained, grown and is able to function outside the body.

An extra ear is presently being constructed on my forearm: a left ear on a left arm: an ear that not only hears but also transmits. A facial feature has been replicated, relocated and will now be rewired for alternate capabilities. Excess skin was created with an implanted skin expander in the forearm. By injecting saline solution into a subcutaneous port, the kidney-shaped silicon implant stretched the skin, forming a pocket of excess skin that could be used in surgically constructing the ear. A second surgery inserted a Medpor scaffold with the skin suctioned over it. Tissue cells grow into the porous scaffold fusing it to the skin and fixing it in place. At present it is only a relief of an ear. The third surgical procedure will lift the helix of the ear, construct a soft earlobe and inject stem cells to grow even better definition. The final procedure will implant a miniature microphone that, connected with a Bluetooth transmitter, will enable a wireless connection to the Internet, making the ear a remote listening device for people in other places. This additional and enabled *Ear on Arm*[9] effectively becomes an Internet organ for the body – a publicly accessible, mobile acoustical organ.

9 *Ear on Arm*: surgical team: Malcolm A. Lesavoy, MD, Sean Bidic, MD and J. William Futrell, MD; stem cell and surgical consultant: Ramon Llull, MD; project coordination: Jeremy Taylor, October Films, London; 3D model and animation: the Spatial Information Architecture Lab, RMIT, Melbourne; photographer: Nina Sellars. Funded by Discovery US.

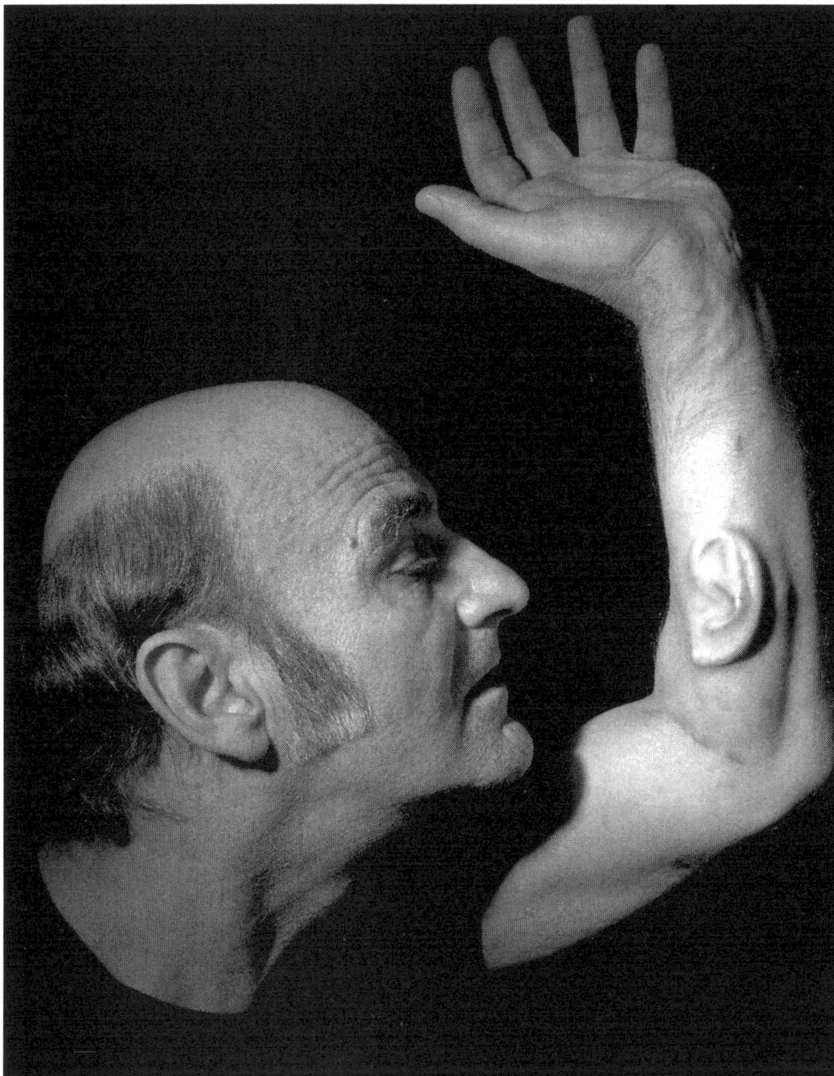

**Figure 6.7** ***Ear on Arm***
*Source*: Stelarc, 2006. London, Los Angeles, Melbourne. Photographer: Nina Sellars.

## Interview with Jens Hausser – *Ear on Arm*[10]

The following interview with Jens Hauser was carried out for the 'sk-interfaces' exhibition at the Foundation for Art and Creative Technology (FACT) in Liverpool (1 February–30 March 2008). It further elaborates on the issues and implications of constructing an ear as a permanent attachment to the body. [Figure 6.7]

JH: What does it mean to move from the artistic use of hard prostheses to soft prostheses?

S: Well, you quickly realize that the body is a living system which isn't easy to surgically sculpt! The body needs time to recover from the surgical procedures. There were several problems that occurred including a necrosis during the skin expansion process which necessitated excising it and rotating the position of the ear around the arm. Ironically, this proved to be the original site on which the 3D model and animation was visualized! Anyway, the inner forearm was anatomically a good site for the ear construction. The skin is thin and smooth there, and ergonomically locating it on the inner forearm minimizes the inadvertent knocking or scraping of the ear. During the second procedure a miniature microphone was positioned inside the ear. At the end of the surgery, the inserted microphone was tested successfully. Even with the partial plaster cast, the wrapped arm and the surgeon speaking with his face mask on, the voice was clearly heard and wirelessly transmitted. Unfortunately it had to be removed. The infection caused several weeks later by the implanted microphone was serious. In fact being admitted into ER with an ear infection caused some confusion to the triage nurse who kept wanting to check the ears on the side of my head! The infection resulted in an operation to extract the microphone, to insert additional tubing around the ear, and a week in hospital tethered to an IV drip. To completely eradicate the infection we had to flush the site every hour on the hour during my hospitalization and I was on industrial-strength antibiotics for about three months.

JH: Since the early *Suspension* pieces the stretching of skin in your work stands for stretching the definition of what a body is. Once this notion is stretched, what does growing add?

S: The suspended body was a landscape of stretched skin. The body was seen as a sculptural object. In constructing the ear on the arm, the skin was also stretched, but in preparation for additional surgical reconstructive techniques and finally stem cell growth to give it more pronounced form. To fully realize the external 3D ear structure will require further surgeries to lift the helix of the ear and construct a conch. This will involve an additional Medpor insert and a skin graft. The Medpor implant is a porous, biocompatible polyethylene

10 'Stelarc: *Extra Ear: Ear on Arm*'. Interview by Jens Hauser. First published in J. Hauser (ed.), *sk-Interfaces. Exploding Borders – Creating Membranes in Art, Technology and Society* (Liverpool: Liverpool University Press, 2008), pp. 102–105.

material, with pore sizes ranging from 100–250 micrometers. This can be shaped into several parts and sutured together to form the ear structure. Because it has a pore structure that is interconnected and omnidirectional, it encourages fibrovascular ingrowth, becoming integrated with my arm at the inserted site, not allowing any shifting of the scaffold. We had originally considered mounting the ear scaffold onto a Medpor plate, thinking that this might elevate it more, and position it more robustly to the arm. But this wasn't the case and this solution was abandoned after being tested during surgery. Now, implanting a custom-made silastic ridge along the helical rim would immediately increase helical definition but also would make room for later replacement of that ridge with cartilage grown from my own tissues. The ear lobe will most likely be constructed by creating a cutaneous 'bag' that will be filled with adipo-derived stem cells and mature adipocytes. Such a procedure is not legal in the USA, so it will be done in Europe. It is still somewhat experimental, with no guarantee that the stem cells will grow evenly and smoothly – but it does provide the opportunity of sculpturally growing more parts of the ear … and possibly resulting in a cauliflower ear!

JH: You seem to consciously create a perceptive conflict between the slowness of the biomedical process and the plug-and-play use of the *Extra Ear: Ear on Arm* as an accelerating electronic body extension.

S: Well, this project has been about replicating a bodily structure, relocating it and now re-wiring it for alternate functions. It manifests both a desire to deconstruct our evolutionary architecture and to integrate microminiaturized electronics inside the body. It also sees the body as an extended operational system – extruding its awareness and experience. When the microphone is re-implanted within the ear and connected to a Bluetooth transmitter, sounds the ear 'hears' will be wirelessly transmitted to the Internet. You might be in Paris, logging into my website and listening to what my ear was hearing, for example, in Melbourne. Another alternate functionality, aside from this remote listening, is the idea of the ear as part of an extended and distributed Bluetooth system – where the receiver and speaker are positioned inside my mouth. If you telephone me on your mobile phone I could speak to you through my ear, but I would hear your voice 'inside' my head. If I keep my mouth closed only I will be able to hear your voice. If someone is close to me and I open my mouth, that person will hear the voice of the other coming from this body, as an acoustical presence of another body from somewhere else.

JH: Therefore one would say that this concept of interface as interference is a recurrent theme.

S: We certainly need to undermine the simplistic idea of agency and the individual. This project links in certain ways to my past work. In the performance *Fractal Flesh* my body movement was controlled by people using a touchscreen interface system connected to muscle stimulation equipment which in turn was connected to my body. [Figure 6.8] In 1995, people in the Centre Georges Pompidou in Paris, the Media Lab in Helsinki and attending the 'Doors of

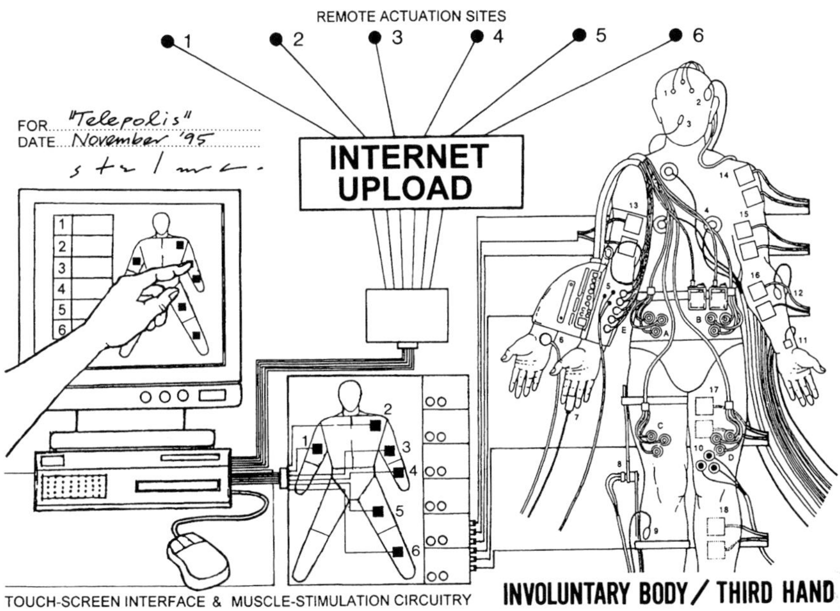


**Figure 6.8** ***Fractal Flesh***

*Source*: Stelarc, 1995. Telepolis, Luxembourg. Diagram: Stelarc.

Perception' conference in Amsterdam remotely choreographed the body (my body), which was located in Luxembourg. Half of this body was controlled by people in other places, the other half could collaborate with local agency. It was a split body experience. Voltage-in (on the left side) moving the body, voltage-out (from the right side) actuating a mechanical *Third Hand.* So the notion of single agency is undermined, or at least made more problematic. The body becomes a nexus or a node of collaborating agents that are not simply separated or excluded because of the boundary of our skin, or having to be in proximity. So we can experience remote bodies, and we can have these remote bodies invading, inhabiting and emanating from the architecture of our bodies, expressed by the movements and sounds prompted by remote agents. What is being generated and experienced is not the biological other – but an excessive technological other, a third other. A remote and phantom presence manifested by a locally-situated body. With the increasing proliferation of haptic devices on the Internet it will be possible to generate more potent phantom presences. Not only is there 'Fractal Flesh', there is now 'Phantom Flesh'.

JH: When Marshall McLuhan devilishly prophesied 'in the electric age we wear all mankind as our skin' he might have referred to connectiveness both as increasing awareness of our media extensions and as a burden as well.

S: It is of course a condition that needs to be managed. What is also interesting is the observation that electronic circuitry becomes our new sensory skin and the 'outering' of our central nervous system. The idea is that technological components effectively become the external organs of the body. Certainly what becomes important now is not merely the body's identity, but its connectivity – not its mobility or location, but its interface. In these projects and performances, a prosthesis is not seen as a sign of *lack* but rather as a symptom of *excess*. As technology proliferates and microminiaturizes it becomes biocompatible in both scale and substance and is incorporated as a component of the body. These prosthetic attachments and implants are not simply replacements for a part of the body that has been traumatized or amputated. They are devices that augment the body's architecture by constructing extended operational systems. The body performs beyond the boundaries of its skin and beyond the local space that it occupies. It can project its physical presence elsewhere. The *Third Hand*, the *Extended Arm* and the *Exoskeleton* walking robot are external machines and systems constructed out of steel, aluminium, motors and pneumatic systems. I was always intrigued by engineering a soft prosthesis using my own skin, as a permanent modification of the body architecture hopefully adjusting its awareness. Of course this architecture is still open to the systemic malfunctioning of the body, the possibility of contamination and infection and to the longevity of the body itself. The biological body is not well *organ-ized*. The body needs to be Internet-enabled in more intimate ways. The *Extra Ear: Ear on Arm* project suggests an alternate anatomical architecture – the engineering of a new organ for the body: an available, accessible and mobile organ for other bodies in other places to locate and listen in to another body elsewhere.

## General comments and concerns

In conclusion, if art is an activity that often deals with the obsessive, the pathological, the perverse (and even the pornographic) and questions our accepted views of aesthetics and ethics, can institutions host such hot, messy and contentious creative activity that goes beyond the constraints and conditions of public and government funding? That is not to say that authentic art practice is only practice that is edgy and destabilizing, shocking or obscene. Perhaps a sophisticated robot might be interesting art; a tissue-engineered object might be perceived as a sculpture; Artificial Intelligence (AI) and Artificial Life (Alife) research can result in intriguing and visually seductive computer animation. It is certainly the case now that much interesting philosophical thinking emanates from new research in cognitive sciences. Artistic practice can be said to be amplified, producing unexpected insights, by the new kinds of imagery that scientific instruments and medical procedures are now able to generate.

The 'Visible Human Project'[11] and Gunther van Hagens' plastinated bodies[12] have spawned multiple ways of viewing and displaying the anatomy of the body. My earlier use of endoscopy (filming three metres of internal space in my stomach, lungs and colon)[13] resulted in the realization of the body as a structure full of spaces that could be physically probed. This led to the *Stomach Sculpture*[14] project in 1993, where the body became a host for a work of art. The *Stomach Sculpture* was not a project sanctioned by a research grant (it was carried out for the Fifth Australian Sculpture Triennale on a modest participation fee augmented by personal funds). *Blender*, a more recent project exhibited in 2006 and in collaboration with another artist, Nina Sellars, involved undergoing surgical procedures to extract biomaterial from each artist's body.[15] [Plate 6.3] The idea was that a machine installation would host and blend the biomaterials for the duration of the exhibition – a machine installation hosting a liquid body composed of biomaterial from two artist's bodies. It was the inverse of *Stomach Sculpture*, where a soft and wet internal body space was the host of a machine choreography. This was self-funded at considerable expense. It was not a project that would have been publicly funded. Although ethical clearance was granted by the University of Western Australia for the *Extra Ear: ¼ Scale*, funding to do the surgical procedures for the *Ear on Arm* project would neither have received ethical clearance nor any institutional or government funding. It was only ultimately realized because of media interest, and funded for a

11 The Visible Human Project® is a dataset of images of anatomically detailed, three-dimensional representations of the human body, developed by the United States National Library of Medicine, <http://www.nlm.nih.gov/research/visible/>.

12 Gunter van Hagen's Bodyworlds, < http://www.bodyworlds.com/en.html>.

13 The three probes into the body were carried out between 1973 and 1975 with the assistance of Dr Mutsu Kitagawa, of the Yaesu Cancer Research Centre in Tokyo. They were originally shot in 16mm film using a gastroscope, a bronchoscope and a colonscope.

14 *Stomach Sculpture* was a self-illuminating, sound-emitting, extending and retracting capsule structure designed as a sculpture for the inside of my body. It was actuated by a servomotor and a logic circuit via a flexidrive cable. It was inserted approximately 40cm into the body. A medical endoscope tracked and video documented the insertion of the sculpture. Fully closed it was 50mm long and 15mm in diameter. Fully open it was 75mm long and 50mm in diameter. It was constructed from titanium, gold and acrylic with the assistance of a jeweller, Jason Patterson, and a micro-surgery instrument maker. *Stomach Sculpture* and the video documentation were exhibited at the National Gallery of Victoria, Australia, 11 September to 24 October 1993.

15 *Blender* was a project co-curated by Kristen Conden and Amelia Douglas for the Teknikunst 05 Contemporary Technology Festival. It was a collaboration with artist Nina Sellars. It was realized with the assistance of Adam Fiannaca (3D modelling and engineering of the installation) and Rainer Linz (Sound Design). The extracted biomaterial was autoclaved to sanitize it and the *Blender* vessel disinfected and hermetically sealed. Both artists were blood tested prior to the surgical procedures. The installation was exhibited at the Meat Market Project Space in Melbourne in 2005 and at the Experimental Art Foundation in Adelaide in 2007.

medical documentary. The *Third Hand*[16] did not receive funding from the Australia Council when it was first conceived and constructed but now such a proposal would probably be successful as a collaborative work carried out at a university. In relation to the *Walking Head* robot, there is an ongoing collaboration at Brunel University, West London between the department of Electronic and Computer Engineering and the Centre for Media Communications Research to improve the design and operation of the present robot and to develop remote actuation from the Internet. This would involve pasting together a choreography from its library of motion icons. The 3D model on the website would simulate the programmed movement, and a second later, somewhere else, the actual robot would move.

In discussing these projects it becomes apparent that artistic practice has to develop strategies in order to interface with the scientific community and academic institutions. It must also incorporate the idea of art as research or establish whether art can flourish at all under the auspices of academic practice. Whether it is meaningful to think of art as research and whether the outcomes are interesting art are open questions. Another contentious issue is the necessity for artists to justify a project within research criteria and scientific narratives to enable adequate funding. For an artist to think about research aims, methodologies and outcomes is to constrain creative thought and practice.

Often what is interesting about artistic practice is what happens between the intention and the final outcome: the slippage between the idea and the actuality. This can happen in artistic practice because the artwork is not primarily concerned with any utilitarian use. There is often no specific goal and no expected results. Art is about play and surprising and unpredictable occurrences, and not about methodical research with particular aims and hoped for outcomes. In peer review feedback from grant applications it becomes apparent that although interdisciplinary research is publicly encouraged, in reality funding bodies and academia still find it difficult to accept the role and the credentials of the artist and to perceive art practice as bona fide within the realm of research. As part of higher education institutions, artists need to cope with the rigid bureaucratic structures and slow procedures. Most of the projects mentioned in this chapter are essentially works in progress or works that will never fully be completed, in that they were initiated with inadequate funding, carried out in limited time and attempted with only some available expertise. This has not been the case for all projects. For example, the *Prosthetic Head* has become a platform for a five-year research programme which includes a number of collaborating universities in Australia and overseas.

16 The *Third Hand* was constructed between 1976 and 1980 in Japan. It was a mechanical human-like hand with an EMG (muscle signal) controlled mechanism with pinch release, grasp-release and 290-degree wrist clockwise and counter-clockwise rotation. It had a tactile sensor system on the fingers to provide a rudimentary sense of touch. It was constructed with aluminium, stainless steel, epoxy resin, acrylic and electronics. It was based on a prototype developed by Professor Ichiro Kato from Waseda University with advice from Professor Shigeo Hirose at Tokyo Institute of Technology, and with engineering by Imasen Denki in Nagoya. It was used for performances from 1980 to 1997.

Being an artist at a university and having a title or wearing a white lab coat neither guarantees interesting art nor meaningful research. Scientists may appreciate artists because of their creative approach. Artists admire scientists because of their specialized knowledge and specific technical skills. We know there are examples of artists doing inadequate science and scientists being naive artists. There are limitations in meaningful collaboration when an artist might not have the mathematical knowledge and language to communicate meaningfully with a physicist. Similarly, a scientist might not have any comprehension of contemporary art practice or familiarity with postmodern discourse. There are times when roles might be happily swapped or confused. For example, during the second operation for *Ear on Arm*, the surgeons constructing the ear on my arm discussed whether they in fact were the artists and the person undergoing the operation was simply the body canvas for their sculpture. There can be intellectual property (IP) issues in a collaborative work, which might extend to the host institution as well as the group of collaborating individuals. So although art as research may have the benefit of access to expertise and equipment and the promise of significant public funding, there are constraints and conditions placed on certain kinds of creative pursuit. This is not in itself a negative aspect, but it most likely excludes much artistic practice as we now know it – constraining the intuitive, the impulsive and the obsessive.

How would universities cope with a Damien Hirst, a Jeff Koons or SRL (Survival Research Laboratories)? Can public institutions cope with performance artists who abuse and mutilate their bodies, indulge in pornographic gestures and carry out violent spectacles? What of university ethics committees or health and safety rules? This is not to say that anything goes, but rather to underline that art can be messy and misunderstood. Art can be disturbing, sometimes dangerous, often perplexing and usually seeks to undermine, expose and amplify. If artistic 'research' is only that which is publicly acceptable, publicly-funded and hosted by institutions then it might not be art at all. Some artistic pursuits are allowed and others are not. Installations such as *Blender*, which involved two artists undergoing surgical procedures to extract biomaterial from their bodies, as discussed, would never have received ethical clearance or funding. The *Ear on Arm* project could only attract funding from the media. Arranging a simple biopsy for a newly-proposed project elicited 51 questions from the ethics committee of one university.

Certainly new media and interactive artworks are producing new kinds of artistic practice that require collaboration and often benefit from industry or institutional funding. In their article 'Interaction in Art and Technology' Linda Candy and Ernest Edmonds assert that 'Collaboration in art practice has grown significantly, in the sense that the visual arts have developed some of the characteristics of film production, with teams of experts working together on projects.'[17] Further, Jean-Paul Fourmentraux says:

17 L. Candy and E. Edmonds, 'Interaction in Art and Technology', *Crossings: eJournal of Art and Technology*, 2/1 (2002), <http://crossings.tcd.ie/issues/2.1/Candy/>.

> The process of technological innovation drives the reorganization of research in the media arts. The imperatives of innovation and creativity have become the driving forces for industry-transferable research and creation. In this context, 'artistic talent' is a highly sought-after, actively encouraged resource, so much so that the identity and role of the contemporary artist are being transformed. No longer only creators, they are expected to be researchers and entrepreneurs, experts in the 'new economy'.[18]

Stephen Wilson takes a positive approach to art as research and makes meaningful distinctions between modernist practice, critical practice and art as research, pointing out that:

> Some artists believe the most powerful response is to become researchers themselves. They attempt to enter into the heart of scientific inquiry and technological innovation to address research agendas ignored by the mainstream and to integrate commentary and play into the research enterprise. I believe this opens up enormous opportunities for the arts.[19]

And undoubtedly, there are artists now who have also been educated as engineers and IT specialists.

At any rate, much art is now conceptually driven and not dependent on the specialist skills of working with one particular medium. The artist is not so much a craftsperson but more like a designer, a film director or an architect who needs to have an overall understanding of a number of related and different skills, but who needs the expertise of others to realize the artwork. Perhaps universities and other institutions hosting artists need to reconsider and reconfigure what it means to do research and to recognize artists' capabilities and credentials for what they are. We should question the agenda of authenticating art practice within the criteria of academic research in universities where the preferred outcome is not an installation or an exhibition but rather a peer-reviewed paper in a prestigious publication.

18 J.-P. Fourmentraux, 'Governing Artistic Innovation', *Leonardo*, 40/5 (2007) 489–492.

19 S. Wilson, 'Myths and Confusions in Thinking about Art/Science/Technology', paper presented at the College Art Association Meetings in NYC (2000).

Chapter 7

# The Garden of Hybrid Delights: Looking at the Intersection of Art, Science and Technology

Gordana Novakovic

My work has formed four distinct cycles.[1] Each cycle is the result of about five years of multi-faceted explorations within the framework of a specific theme and through a variety of media. Although there are notable threads of symbols, phenomena, topics, forms and methods passing through the cycles, each cycle is conceptually cohesive and distinct. I will tackle the role of Information and Communication Technology (ICT) and the evolution of research elements in my artistic practice through a case study treatment of some of my works from the four work cycles. Focusing on questions of artistic method, context, documenting and archiving will allow me to address some broader issues and problems regarding the 'ICT-enabled' arts.

Questioning my artistic method unravels the interplay between artistic, philosophical or scientific components and technological forms (be they algorithms, software architectures, or different output tools) that have inspired me, and the methods of applying different technologies, crafts and materials. One of the key characteristics of ICT-enabled arts is their method of production which I regard as being inseparable from artistic concepts and methods. To focus on production is to call into question the usual categorizations applied to the field. Some of the terms used originated in symposium and festival propositions; others have been promoted by theorists, art historians and curators. My practice, which intersects art, science and technology now reads almost like a brief history of such categorizations under the broad, quite unfortunate, title 'New Media': mixed media, multimedia, computer art, computer animation, net.art, interactive art, etc.

Two of my four work cycles were developed during the collapse of Yugoslavia. Belgrade, where I worked, was only 100km from the battlefields. Although my works do not address the war explicitly, many were influenced and conditioned by it. To put the Belgrade works and experiences side by side with those produced in my last ten years working in London puts the themes of artistic method, context, documenting and archiving in clear relief.

1 <http://www.gordananovakovic.net/> (website updated 2009). (All URLs current at the time of writing.)

Art is of course inherently political and influenced by cultural and socio-political context, but technology-enabled artworks are also conditioned by the economy, and by access to advanced technology and scientific research. At the turn of the century, these arts are becoming even more the privilege of the wealthy, both nationally and internationally. Another set of problems in the ICT-enabled arts stems from the mixing of media and art disciplines, the real with the virtual, in the aesthetics of non-linear processes and computational models. From the aesthetics of large-scale spectacular interventions in landscape to the aesthetics of nano-scale constructions, these artworks collide with the fundamental methodologies of art history, archiving and museology.

## The background: Learning by playing

In the late 1970s I was in the process of confronting the practical problem of rendering tangible (in the form of a painting or drawing) the mysterious interplay between conscious, cerebral processes and irrational, unconscious, emotional ones. What if one were to use a computer as a translating or decoding machine? Could it work as a tool facilitating communication between these different realms? Juxtaposing a computer-generated image with painted surfaces led me to start playing with and learning from the first generation of IBM personal computers in 1984, and shaped and informed my interests in psychology and philosophy. To produce an image or a sound that is in its essence mathematical, and which results from computational processes, is a common conceptual framework for a large number of computer artists, and I too am interested in the aesthetics of this abstraction. But it is not the aesthetics of the visual output of an algorithm, or the possibility of mechanical art that fascinates me; rather, it is the hybrid that arises from the powerful tensions between software structure and fine art aesthetics. Assisted by my mathematician friends I worked with 3D topographic imaging software designed for geologists, to produce an algorithm that was to become a leitmotif of my works. Like Frank Stella (whose work I was unaware of at the time), I experimented with pen plotters, photo-enlarging the images generated by the algorithm to make silk-screen prints, and combining them with traditional fine arts techniques. [Plate 7.1] The enlarged computer-generated patterns produced ambiguous responses from viewers. They were often mistaken for painstaking, precise hand-drawings, and the demystification was often a disappointment. For fellow artists, let alone art theorists and critics, it was not art at all. The work of groundbreaking Yugoslavian artists from the 1960s such as Vladimir Bonačić, Petar Milojević and Tomislav Mikulić seemed to have left no trace. (Since the early 1970s, most of these, including Edvard Zajec, perhaps the best known internationally, had transferred their activities to Western Europe and the United States.) The early works of these pioneers were mostly influenced by cybernetics; in contrast, I had no intention of introducing such a discourse – I was motivated by the pure excitement of experimenting with new tools. Questions of discourse and historiography would come later, with my search for an audience.

## The first cycle

### *Parallel worlds (1988–91)*

During the years of soft communism, culture and the arts in Yugoslavia were strictly governed by national institutions, and reasonably well funded. There was no art market, but there was a tradition of intense exchange with the international avant-garde. Between 1961 and 1973 the Contemporary Art Gallery (Zagreb) hosted five international exhibitions and symposia entitled 'New Tendencies', dedicated to 'visual research', with over 300 participants from the fields of computer art, cybernetics and 'art and science'. In 1972 a cluster of experimental activity took place at Elektronski studio, established by Radio Beograd, two years after Pierre Boulez's IRCAM was founded. When, in the 1980s, the IBM PC catalysed the global emergence of 'electronic art', soon to become 'New Media', there was keen interest in such works which were being shown at newly-established international festivals where the ethos was open and non-institutionalized. Art theorists and curators devoted to the field were rare, so as well as producing artworks, artists created the theoretical foundations for the discipline through writing, curating events, and building common or personal archives. The interest, involvement and influence of academia was minimal. Ars Electronica was not yet an expensive spectacle. In the 1980s Yugoslavia entered the post-Tito era, marked by progressive economic decline. Artistic experimentation with technology was mostly 'garage production', and the erosion of scientific and cultural investment pushed experimentation and research away from institutions into informal cross-disciplinary networks attracting artists, scientists, mathematicians and engineers.

Marjan Šijanec, one of a handful of composers of computer and electro-acoustic music in Belgrade, worked with me on *Parallel Worlds*, our first video piece; it was based on my works on paper, which mixed silk-screen printed algorithms with pastel, watercolour and China ink. [Figure 7.1] The sound was an interplay between electronic music and live piano improvisation. To explore the mixed media framework we set up a structure for recombining audio-visual modules consisting of electro-acoustic music, computer music, live piano improvisations, mixed-media paintings, and video. Within the concept of juxtaposing the traditional media of fine arts and music with contemporary electronic media, we questioned the changes in perception and aesthetics formed by technology. In the catalogue for *Parallel Worlds*, I wrote:

> We are living in a time when the borders between the spheres of action of the human mind are disappearing. [...] Visual arts, thanks to computers, enter the domains of electronics and science in general. Art, using new technologies, changes in accordance with altered human environments. We are bombarded, simultaneously, with various sensations [...] surrounded by TV screens, monitors, radio waves etc. [...] [W]e often perceive all these sensations

> simultaneously, completely accustomed to this type of 'future shock'. The result is a new aesthetic [...][2]

To test our concept, we put on live performances with accompanying video screenings, electronic sound and live improvisations. This work was presented at a number of experimental music events and solo multimedia exhibitions.

**Figure 7.1** ***Parallel Worlds*, 1990**

*Note:* China ink, watercolour and silk-screen print.

*Source*: Used with permission of G. Novakovic and A. Zlatanović.

In May 1991, with colleagues from Novi Sad, Marjan Šijanec and I organized the Exhibition of Yugoslav Computer Art in the ULUS Gallery in Belgrade. Everything was installed using the artists' own equipment. The exhibition featured a variety of approaches to computer technology, spanning mixed media objects, live computer music concerts and happenings with authors from Slovenia and Serbia. Our aim was to reignite the field and to present the computer as an artistic tool and medium. The venue was prominent, and the show received interest and acceptance from the general public. Riding this wave, we next established the Association of Electronic

2 G. Novakovic and M. Savić, *Under the Shirt of a Happy Man* (Belgrade: Soros Foundation, Department for Contemporary Art, Belgrade, 1995), p. 2.

Media Arts (AUEM). Most of the eighteen founding members are now academics, such as Dr Miško Šuvaković, Professor Čedomir Vasić and Professor Vladan Radovanović, who have continued to contribute to the field through their artistic practice and theory. Our exhibition took place just one month before Slovenia and Croatia declared independence, which led to the escalation into full-scale war in July of the same year. This was followed by the United Nations sanctions against Serbia and Montenegro, the only republics remaining in Yugoslavia, for fomenting war in Bosnia and Croatia. I have no knowledge of what happened to the video documentation of the exhibition in the subsequent confusion; the only legacy is a modest exhibition catalogue with a short text by Predrag Šidjanin.

## The second cycle

### *The Shirt of a Happy Man (1991–5)*

After my initial computer explorations, I named the recurring algorithm *Form*: a symbol and image, a set of coordinates. The computer animation *The Shirt of a Happy Man* (1991) was a contribution to the quest for the Holy Grail of computer art: generating sound from image and vice versa. The composer Miroslav Savić and I devised a process for decoding sounds from basic wireframe prints of the algorithm. [Figure 7.2] His musical score determined the dynamics, rhythm and duration of the computer animation. The post-production was low-key: we were permitted to occupy a television studio after work hours (from 10.00 pm to 8.00 am). In return we developed the same computer-generated material into a series of advertisements for the weekly English edition of a local daily newspaper. Inspired

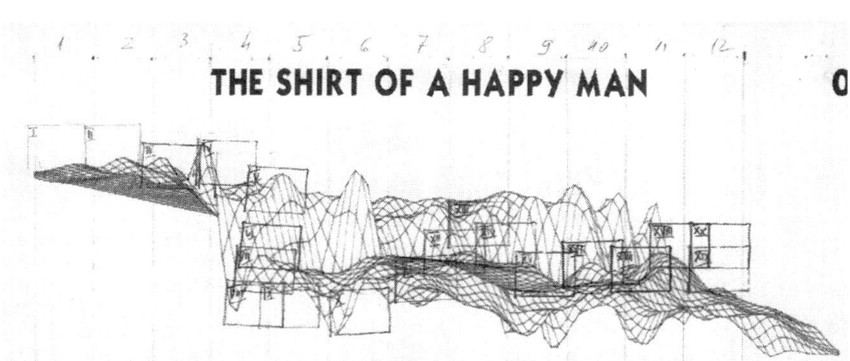


**Figure 7.2 The representation of the algorithm *Form* that served as an abstract script and musical score for the computer animation *The Shirt of a Happy Man*, 1991**

*Source*: Used with permission of G. Novakovic.

by Miroslav's interest in theories about the relationship between colours and numbers, and my interest in C.G. Jung, we wrote: 'This work is a parable about the endeavour to achieve complete harmony within one's own personality. In this endeavour, *Form* experiences certain transformations, thus achieving the essential unity of music and painting.'[3]

There is something puzzling about plotters. Their human-like movements strike me as performative: executing strokes while pausing to calculate the next move for a few seconds, only to continue working on another, often distant section of the paper's surface. In *Plotter Form* (1993) the algorithm was simultaneously treated as a visual and musical 'score', as well as being the source of the set of movements of the output mechanism (i.e. the plotter). Miroslav Savić's musical composition, extracted from the elements of *Form*, was combined with my painting of an abstract dreamscape, and offered as a 'canvas' to the plotter. The oil on foil painting was inserted into the plotter with an audiotape ribbon of Miroslav Savić's music attached along one of its edges. As it moved the foil backwards and forwards in drawing the algorithm, the plotter simultaneously pulled the tape over a tape head attached to the side of its casing, playing sequences of music backwards and forwards. The sound of the plotter's motor and the strokes of the pen combined with the music. The Holy Grail was replaced by the *deus ex machina*.

Unfortunately, the video documentation of the performance was later erased from the master tape by accident. This bitter lesson made me think about the significance of documenting this kind of art, and about its transient nature. So, what if we turned the video documentation, the account of an artwork's exhibition, into a means of artistic expression, thereby producing a hybrid? With this idea in mind we carefully planned the scenario for documenting *White Shirt* (1993), a performance undertaken in collaboration with the experimental dance group Mimart. I wrote:

> The mise-en-scene for the performance is an installation, consisting of a 100m long fishing net and ship-ropes. It is overlaid with a video projection of computer-animated sequences shown in a repetitive mantric cycle. The elements for the animation are the sphere and circle representing universal cosmic forms. The performers inter-act with the environment/installation and the computer-generated animation sequences.[4]

The synergy of the different artists and forms culminated in an installation space which created tension between the body and digital technology, providing a stage for the performers' visceral, intuitive responses. The movie was part document of the event, and part video dance, with 3D animation processed from my images and superimposed onto the raw footage of the performers. [Figure 7.3] Some simple ready-made software was used to animate my ink-on-paper drawings of

3 Ibid., p. 7.

4 Ibid., p. 11.

**Figure 7.3 A still from the video of *White Shirt*, 1993**

*Source*: Used with permission of G. Novakovic, M. Ristić and R. Novakovic.

the spherical, cell-like forms. The result was premiered at the second ISEA festival in Helsinki in 1995, and also at a number of specialized video dance festivals. It also features in the archive of the Mimart group, which still exists.

In 1993, the Soros Foundation became active in Serbia and other parts of former Yugoslavia. It supported the founding of Opennet, Belgrade's first Internet Service Provider. From the perspective of the cultural ghetto produced by the 1992 UN sanctions, the potential of this new platform was more than obvious, and in parallel with my other activities, I worked on my own website in the form of an online catalogue. I was one of the few artists invited to contribute to the foundation of Cyber-Rex, the first new media centre in Serbia. Rex soon became a meeting point for a number of young designers, architects and digital artists, as well as the artistic association Apsolutno (Zoran Pantelić, Dragan Rakić, Bojana

Petrić and Dragan Miletić) and the Škart group of designers (Dragan Protić and Đorde Balmazović). Rex also hosted political discussions and fora, as well as exhibitions and workshops with young practitioners, activists, and organizations from former Yugoslav countries (such as the Croatian magazine Arkzin), and talks with international speakers such as Stephen Kovats, director of the International video forum Ostranenie.

In the late 1980s interactivity had emerged as a groundbreaking concept. Interactive art still remains the newest artistic form, fundamentally different from the artforms practiced for thousands of years. It questions the role of the author, while the viewer becomes an active, physically engaged participant in the process of the unfolding of the artwork. An interactive artwork is a technology-enabled, real-time processed blend of audiovisual, kinaesthetic and haptic experiences. It seduces and almost overpowers the senses; it is a dialogue, but also a confrontation between the human body and technology.

During the early years of sanctions, war and economic collapse, I read about this blossoming field in scarce catalogues. Fascinated by its potential, in 1994 Miroslav Savić and I, joined by electronic engineer and software designer Zoran Milković, started work on *Under the Shirt of a Happy Man*, which would be the first interactive installation in Serbia. The installation consisted of an exercise bicycle, a fixed video camera and a composite audio/video projection controlled by a PC network. A pulse reader input the cyclist's heart-rate, changing the audio/visual parameters in real time. The projection faced the participant and consisted of a real-time composite image of the cyclist, an oil-on-canvas dreamscape and the shifting wire-frame of the algorithm. The use of the bicycle was quasi-functional, and differed conceptually from Jeffrey Shaw's *The Legible City* (1989–1991) in which the handlebar and pedals provided the viewer with control over direction and speed of travel in the virtual world. In contrast, the exercise bicycle did not influence the work directly; its function was to provide an everyday kinaesthetic space and to influence the participant's heart rate. The piece utilized home equipment: a local PC network consisted of our own computers, my home video camera, a borrowed exercise bike and private editing studios. In the catalogue we wrote:

> The basic concept lies in the correlation between the spheres of the conscious and the unconscious, creating a juxtaposition of two parallel visual currents. The idea is to merge the psychological and psychosomatic processes with visual and aural material expressing the unconscious worlds of the authors. The participant is not in a position to consciously control the visual and aural changes; thus, the act of experiencing the offered artistic material is directly (unconsciously) communicated to the viewer, jointly creating a novel value, instantaneous and unique.[5]

5 Ibid., p. 15.

In my continuing search for an 'ideal' form of interactive installation, the concept of intuitive interaction is both the start and end point in developing a piece. This remains a major principle of my artistic practice. *Under the Shirt of a Happy Man* was exhibited in 1996 as a solo show and also as an invited part of BITEF, Belgrade's international festival of experimental theatre. [Figure 7.4] This spontaneous recognition from the world of theatre inspired my later thinking on the theatrical and ritualistic elements of interactive installation. However, it was in the form and the mode of production that I first found the most obvious commonalities between interactive installation and theatre, through comparing the experience of 'performing' the same piece in a fine art gallery, and as an experimental theatre performance. A contemporary theatre offers flexible use of space, basic equipment, and, more importantly – dark space. What might seem to be trivia are in fact some of the major production constraints in the practice of interactive art.

**Figure 7.4 A still from the video-documentary *Under the Shirt of a Happy Man*, the first interactive installation in Serbia, 1993**

*Source*: Used with permission of G. Novakovic and P. Popović.

## The third cycle

### *Infonoise*

The Balkan leaders met at Dayton on 21 November 1995. Although they agreed on a comprehensive settlement to the 43-month long Bosnian war, the late 1990s in Belgrade were overshadowed by new conflicts. Between 1993 and 1997 ethnic tensions and armed unrest escalated, eventually leading to the war in Kosovo. In 1997 Ars Electronica announced the theme of its coming festival: *Infowar*. This topic resonated strikingly with the context of Belgrade, where the abstract topic captured everyday life, saturated with manipulated disinformation, warmongering, banality and spectacularization of tragedy. One was confronted with the nature of modern warfare, which eliminates the difference between the civilian and the soldier, and kills the truth by means of the media, which binds with the military into an ICT-enabled war-machine. I began to explore the phenomenon of information noise and the repetitive nature of media war propaganda, infowar, and the specific conjunction between media, politics and the military complex. Extensive research in the archives of Belgrade daily newspapers resulted in a minimalist bilingual net. art *Info-Noise* piece in June 1998. Cut-ups of daily news headlines turned into a noise poem, an instant expression of my state of mind at the time.[6]

In our continuing explorations, Miroslav Savić, Zoran Milković and I aimed to create a platform that would engage a participant in spontaneous, non-verbal dialogue with an autonomous virtual organism. I wanted to move away from the rectangular, essentially cinematic, projection screen and the passive viewing experience towards a setting enabling active participation and kinaesthetic experience. In my search for theatrical and sculptural elements I linked two symbols: the Ouroboros and the Möbius strip.[7] These symbols relate strongly through the meanings they engender and their representations; one is archetypal, and the other mathematical. We compensated for our inadequate resources with creativity and decided to conduct the experiment in virtual space first, by creating an animated three-dimensional computational model; the method was more scientific than artistic.

Developing a piece through prototypes, whether computational or physical models, stimulates cross-disciplinary synergy, but also allows the material properties of hardware and software to assert themselves in the creative process. This became one of the foundations of my future working method. I worked with Zoran Milković, and with Milena Mandić, an architect and expert in 3D graphics. We met almost on a daily basis for several months, working on the animation and learning from each other. I gained an understanding of the aesthetics and basic

6 G. Novakovic, 'INFOWAR: Info Noise in Belgrade – part 1', INFOWAR Ars Electronica 98 Linz Austria netsymposium discussion, <http://www.aec.at/infowar/NETSYMPOSIUM/ARCH-EN/msg00144.html> (website updated 9 June 1998).

7 A. Guaridis, Infonoise, INFONOISE Interactive Gallery Installation and Web-Connected Theatre Event, <http://www.infonoise.net/news/Angueliki.html> (website updated 2002).

properties of computational processes and the tools involved, but also contributed my knowledge and experience as a trained painter. Incorporating the attributes of traditional visual aesthetics by using the visual art techniques of perspective, colour, and sfumato to suggest depth, diminished the crudeness of the computer-generated image. [Plate 7.2] It was a collective painting with technology.

The original concept of an on-site interactive installation then expanded into an online event, the virtual *Theatre of Infonoise*. This was organized in emulation of new media centres, such as the Institute of Contemporary Arts in London, or Cinema Rex in Belgrade (our host institution). It included a foyer, experimental theatre space, exhibiting space, café and bookshop. The virtual theatre itself was constructed as an abstract stage following the model of a theatre-in-the-round. [Figure 7.5] In the planned final installation, the sculptural on-site Möbius strip and its online mirror image were designed to be connected in a feedback loop. Online participants were to contribute local headlines from different parts of the world which would form the textual modules. The arrangement and dynamics of the headline modules would mirror the trajectories of participants within the physical installation, and were designed to be featured on both the virtual stage and the on-site installation once we moved from the 3D model to the actual production.

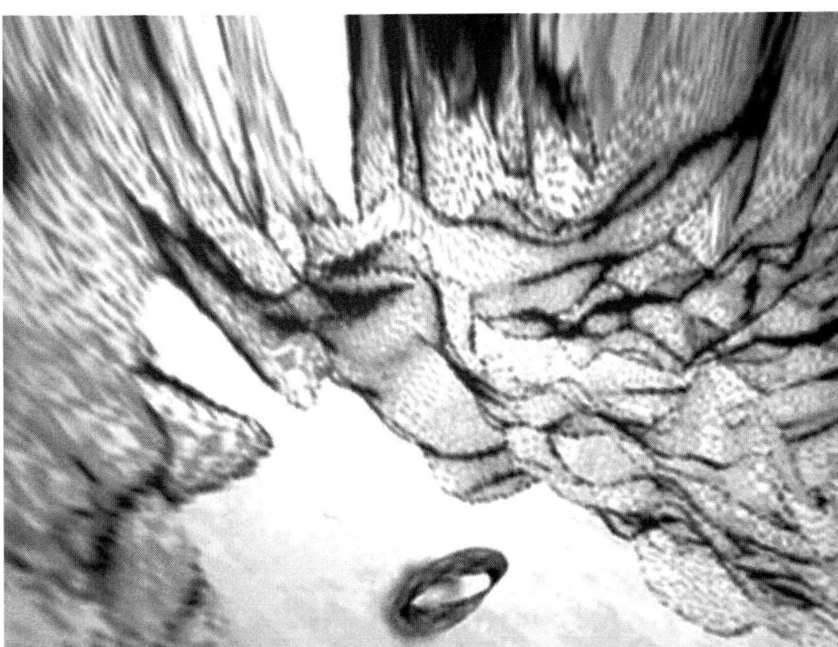

**Figure 7.5 A still of the abstract stage and auditorium from the virtual (3D computer simulation) *Theatre of Infonoise*, 1998**

*Source*: Used with permission of G. Novakovic and Z. Milković.

In June 1998 Miroslav Savić and I organized a five-day symposium entitled 'A Short History of Electronic Art in (Former) Yugoslavia, Part One', supported by the Cinema Rex Centre. We invited all seventeen founding members of the short-lived AUEM Association to present their current and past works. The event attracted a cross-generational audience of practitioners from the field, and a handful of connoisseurs. The aim was to establish a foundation for documenting and archiving electronic/computer art. The 'Part One' in the title of the symposium underlined our ambition to begin a programme of rigorous historical research. The symposium coincided in time and intention with 'Dialogues with the Machine', held at the ICA to mark the thirtieth anniversary of 'Cybernetic Serendipity', the 1968 exhibition of computer art at the Institute for Contemporary Arts, in London. A number of videotapes covering our symposium are, I hope, still neatly packed somewhere in Miroslav Savić's house. However, the real-time online coverage of the event, with video clips, was lost somewhere in the turbulent history of Cinema Rex. To date there has been no further interest in continuing this initiative. The manifold consequences of the civil war are still very present in all areas of everyday life, and the concept of a 'Yugoslav' pre-war cultural scene is in opposition to the agenda of promoting newly established identities, and enforcing national cultures, history and languages.

Looking back on this phase, ICT showed itself to be a potent tool, especially in dramatic socio-political contexts where online communication is a necessity, and documenting and archiving are difficult. The Internet continued to be a vital means of breaking through the isolation produced by the 1992 international sanctions to 'suspend scientific and technical cooperation and cultural exchanges and visits involving persons or groups officially sponsored by or representing the Federal Republic of Yugoslavia (Serbia and Montenegro)'.[8] The website and personal digital archive that I had started in 1993 were not an option, but had been effectively enforced by the circumstances. In 1998, when I made the move to London that I had planned for two years, I faced the challenge of packing 20 years of work in a small suitcase. This inevitably resulted in the loss of a large number of artefacts, but the experience of sifting and parsing my past also provided a framework for the documenting and archiving which continues to be an integral part of my artistic practice. The disadvantages of digital archives stem from the nature of ICT: its rapid development quickly turns what is state-of-the-art into junk, and necessitates the continuous work of converting material to new formats. This leads to a process of constant consolidation between processes, ideas and blind alleys that may be several decades apart. On the other hand, in my multidisciplinary teams, the online collaboration and the process of translating ideas and processes from one discipline and technical language into another is instantly suited to website presentation, enabling us both to chart and to strengthen our working methods.[9]

---

8 UN Security Council 3082nd Meeting Resolution S/RES/757 30 May 1992.

9 G. Novakovic, R. Linz and Z. Milkovic, INFONOISE interactive gallery installation and web-connected theatre event, <http://www.infonoise.net> (website updated 2002); Novakovic, 'INFOWAR'.

My move to London coincided with the dramatic escalation of the Kosovo conflict. Moving from the underground works made with friends in my native city, which were inevitably embedded in its chaotic socio-political landscape, I now faced the complex fabric of London. I found a profit-driven art world, mirrored in a low-key experimental scene still coloured by late 1980s media activism, and an emerging niche for New Media within academia. In 1999 Infonoise was scheduled for exhibition by Benjamin Weil, then Director of ICA New Media. However, Weil resigned due to lack of funds, and the exhibition was cancelled. Shortly afterwards, Infonoise expanded into a less localized concept through my links with Rainer Linz, the Australian composer of new music. We were introduced in 1999 at the Festival of Computer Art in Slovenia, and a few months later we started an online collaboration that still continues. Within my artistic method, the influence and contributions of my close collaborators' personal philosophies form my works to a great extent. Rainer Linz's explorations of the phenomenon of noise and the physiological impacts of music have been invaluable, and his introduction of the framework of polyphonic music has led to a major conceptual shift in my work.

The concept of polyphony inspired our performance-presentation at ISEA 2000 in Paris, that played with transforming a cliché into an artistic form. A computer animation based on the movements of the Ouroboros was accompanied by a pre-recorded soundtrack of Rainer Linz's and Zoran Milković's reading (in English and French) of the project outline. My live delivery of the same text in Serbian was mixed with theirs and with the Ouroboros soundscape. The voices were structured in the form of a musical canon. The artistic statement and technical descriptions became a trilingual poem, and the conference presentation turned into a performance.

In May 2001, after six years of work on Infonoise, we were offered a week to install and exhibit the full-scale installation in Cinema Rex. We jumped at the chance. The team met for the first time after almost two years of online collaboration between Belgrade, London and Melbourne. The major task we faced was the production of the monumental, 12m by 1.50m Möbius Strip. Extensive consultations with university experts produced suggestions for complicated, costly, heavy and robust solutions based on bulky supporting structures. We decided to take our chance with a sculptor, working with polyesters, who agreed to do it our way: that is, in the gallery space, on the fly, joining the ends of the 12m sheet to form a light and smooth Möbius Strip which was suspended by fishing line. [Figure 7.6] It was a crazy idea – and certainly would not now pass health and safety regulations as it produced lots of unbearable toxic fumes. Using our 3D computer simulation as a guide, we started the painstaking work of assembling the hardware (a twelve-PC network, Cinema Rex's support in kind) and tuning the software. We opened the gallery space to the public for four hours a day, allowing them to follow our progress and familiarize themselves with the artwork. The work was finished at the last moment in time for a live video-stream between Belgrade, Paris and Marseilles. It was simultaneously the opening and closing night. During the process of building the installation, the dialogue with the

**Figure 7.6 A still from the video-documentary of Infonoise, showing the positioning of the Möbius Strip**

*Note:* The projection, sound, and interaction were controlled by the row of computers in the background, 1998.

*Source*: Used with permission of G. Novakovic and N. Majdak.

general public was invaluable. It gave us the opportunity to nurture an informed and engaged audience which led to their giving a spontaneous visceral response to the work, based on their familiarity with its conceptual and technological complexities. A number of participants showed signs of full immersion, characteristic of VR environments. This unanticipated response inspired further research in the fields of theatre[10] and psychology, and eventually led me, quite logically, to neuroscience, my current interest.

10 G. Novakovic, 'Electronic Cruelty', in R. Ascott (ed.), *Engineering Nature: Art and Consciousness in the Post-biological Era* (Bristol: Intellect, 2006).

## The fourth cycle

### *Fugue*

Exposure to the overwhelming environment of metropolitan London, saturated with technology, inspired my exploration of a number of different fields. I was looking for the key to understand how perceptual and cognitive processes changed through their interaction with digital technology. I was also interested in assessing the contemporary role and impact of ICT on many aspects of human well-being. One of the most significant changes in my circumstances was the opportunity to access advanced scientific research and technologies in London. This road led me to the university system, and to the concept of research as artistic practice. An Arts Council England Individual Grant in 2004 brought modest funding for *Algorithmica*, originally titled *City Portrait*. Firmly grounded in research, it aimed to address critically the form of the mass-media industry. It was a spontaneous, non-tactile interaction, based on the biological principles of interaction among cells. It applied a game software architecture within the setting of a fictional 3D London tube map. The piece was conceived as using multi-platform virtual reality technologies. The project changed when Dr Peter Bentley, of the Department of Computer Science at University College London, joined the project. Dr Bentley works on computational models of the human immune system. We formed a team with Rainer Linz, and with Anthony Ruto, a PhD candidate in 3D modelling from the same department. Although we had access to the virtual reality facilities at UCL, we decided to abandon the idea of using VR and CAVE technology, finding its numerous limitations to be in clear opposition to our concept of a spontaneous, intuitive interaction with the general public.

The year 2004 brought an AHRB/ACE Art and Science Fellowship, and a Leverhulme Trust artist-in-residence award. *Algorithmica* evolved into *Fugue*, a project with two potentially conflicting goals: creating an artwork, and developing an audiovisualization for scientists. Our ambition was to explore the possibilities of designing a system for calibrating interaction through cross-disciplinary research in the field of perception – a 'tuneable' scientific tool/artwork. Our focus was to understand and apply the principles of biological processes rather than to create photo-realistic 'beautiful imagery', or merely re-represent scientific findings as visualizations or sonifications. The immune system soon became our main interest. In a five-year programme of scientific research, Dr Bentley had produced a computational model of the immune system. I had become fascinated by the images of immune system cells prepared by Dr Julie McLeod, of the School of Life Sciences, University of the West of England. My friend, the immunologist Dr Nada Pejnović, a research fellow at the William Harvey Research Institute, London, gave me a detailed introduction to the field, and I was able to discuss with her my artistic interpretations of scientific subjects, my sketches inspired by medical books, and the concept for the artwork. Real-time generated images originating in computational processes would set the framework for the visuals, with a major practical requirement being to reduce the typically heavy computation to a minimum. To meet this condition, and to achieve an abstract, symbolic representation of the actors in the immune system drama, I looked back to the cell-like, egg-shaped and spherical structures that appeared in my paintings from the 1990s and suggested making some clay models, as I had done for my early paintings. Anthony Ruto scanned the clay models in 3D as the starting point for the final look of the inhabitants of our virtual immune system. To emphasize the focus on processes and the distinction from the aesthetics of the gaming industry and commercial computer graphics, I suggested a black-and-white approach. This would also assist the future development of the science-based methodology for analysing calibration. Both scientists found this idea problematic, because, in their own words, they 'could think of the immune system only in red'. As a compromise, our first prototype was indeed monochromatic, but in red.

Anthony Ruto and I met on a weekly basis for a couple of months. The combination of his expertise in creating 3D wire-frame models of the human body, his taste for abstract visual art, and my experience, gave rise to an enjoyable creative process. We replaced the monochrome red colour with greyscale, which was now accepted as being congruent with the overall conceptual framework. Rainer Linz designed the sound software around a series of customized audio players that he called *Fugue Players*, which responded in real time to changes within the artificial immune system. The first outcome of our collaboration was a conceptual paper presented at the fourth International Conference on Artificial Immune Systems, in Banff, Canada, in 2005.[11] Howevei, our concept received little attention, as most of the scientists categorized it as 'non-scientific', although a small number praised its fresh approach. It was only as this text went to press, that we noticed some slight interest from the community dealing with scientific representations. This seems to be a rather static field, despite the fact that multimedia tools and technologies have been available for decades.

After the paper, we concentrated on developing the artwork. At the time, I wrote:

> The title *Fugue* is a metaphor for the transdisciplinary nature of the work, and for the method applied: inter-weaving the different perspectives of artists and scientists. The emergent, evolving nature of the artificial immune system algorithm, the use of repetition in the form of a succession of variations of 'events', and the complex structural and functional interrelationships between the individual elements and processes are strongly related to the musical form of counterpoint, which formed one of the inspirations for the artistic concept for *Fugue*.[12]

11 <http://oak.cpsc.ucalgary.ca/ICARIS-2005/CFP.html>.

12 P.J. Bentley, G. Novakovic and A. Ruto, '*Fugue*: an Interactive Immersive Audiovisualisation and Artwork using an Artificial Immune System', in *Proceedings of ICARIS 2005, Artificial Immune Systems* (Berlin/Heidelberg: Springer, 2005), p. 4.

We were engaged in intense online work, shaping the architecture and aesthetics of the interactive installation. Online communication imposes numerous limitations and disadvantages, from the inevitable time lags to the lack of face-to-face discussion. After a period of excitement and enthusiasm, we entered a phase where tensions ran high. This was caused by the incompatibility between scientific accuracy and the artistic interpretation of scientific data, and it escalated to a level that almost threatened to end the collaboration and the project. It took a lot of effort from all of us to reach a consensus, and we decided from then on to keep our comments within our own area of expertise.

Rainer Linz and I wanted to provide a setting for a dialogue between the rhythms of the piece and the biological rhythms of the participants – between the computational model and the living body. In a visit to Belgrade, I discussed the sculptural form for the set with Zoran Milković, and the outcome was a structural concept reminiscent of a lymph vessel. This brought a much-needed new impetus to the project. One of our major goals was to minimize the production costs and to simplify, as far as possible, the setting up of the piece. Zoran Milković again produced a 3D computational model that served as a virtual maquette for designing the interaction software, and also as a manual for the final production. Richard Newcombe, at that time a researcher at University of Essex, and now a PhD candidate in the Department of Computing at Imperial College, London, joined the project, and brought a new layer of complexity to the software architecture. For the interaction between the sound and the graphics, and between the participants and the installation, he adapted his particle filtering and computer vision software, originally designed for scientific research, to meet the needs of the *Fugue* installation. The University of Essex provided access to a large arena (used for robotics research), and also gave support for building a full-scale mock-up for testing and finalizing the software design. [Figure 7.7] After a few more weeks of exciting, intense work on the project, developing a common language with a new collaborating scientist in the process, we had our piece fully developed and tested. The first large-scale interactive *Fugue* installation took place in Belgrade in August 2006, again in the prominent ULUS Gallery. [Figure 7.8; Plates 7.3 and 7.4] *Fugue* had been selected for exhibition in the gallery through a standard annual competition, and as part of the award received some modest funding for the production costs. However, during the selection process its distinct scientific emphasis met with strong antagonism from some members of the all-artist jury.

The *Fugue* experience shed light on some of the problems of any artistic practice that falls within the intersection of art, science and technology. The production and exhibition phases have still not been recognized as an integral part of research in the arts. Current curatorial practice and exhibition policies still present problems for work in this area, even at international level. Another set of difficulties can be traced to the absence of funding for a substantial commitment from scientists, and to the lack of acceptance of the validity of these collaborations within the scientific community. At present, in contrast to the artists involved, most of the scientists work on these projects out of personal interest and in their limited spare time. This

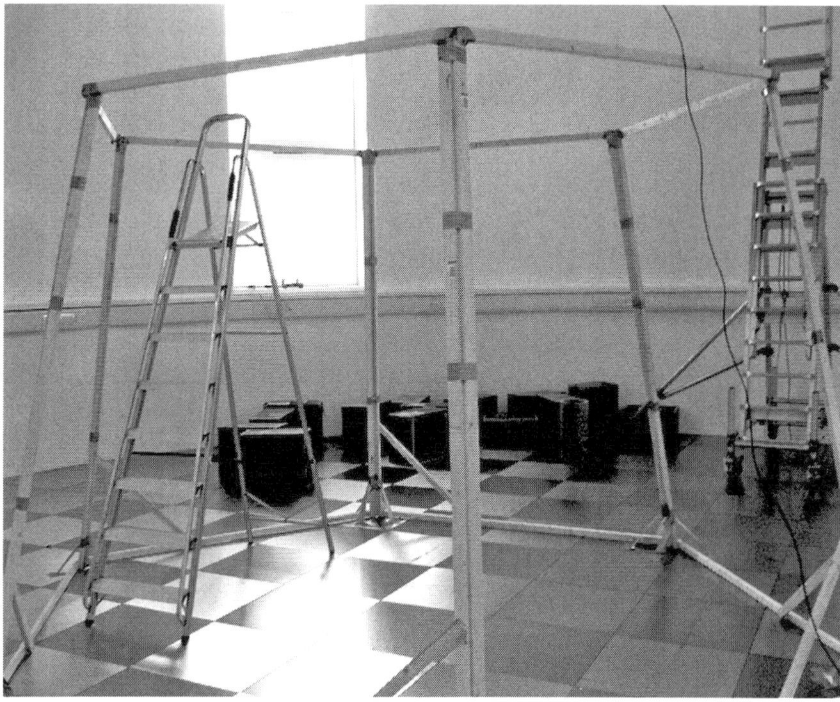

**Figure 7.7 The full-scale mock-up of the *Fugue* installation built at the University of Essex for developing the interactive software**

*Source*: Used with permission of G. Novakovic.

often results in an imbalance between the artistic and the scientific contributions to these hybrid projects. Consequently, the field remains unstable, in a kind of no man's land between the art world, the humanities and the sciences. The road to an exhibition is exciting and inspiring, but also long and troublesome.

## The Tesla art and science forum

My role as the first artist to take up a residency at the UCL Computer Science Department was to introduce over 60 academics and researchers to the concept of art and science, to engage them with this area, and to create a context for my work within this new environment. My experiences in this position have given me further insight into research-based practice in a university environment. With my close collaborators on the *Fugue* project, I decided to start a Research Interest Group in the Department,

**Figure 7.8 Building the *Fugue* interactive installation in the ULUS Gallery, Belgrade, 2001**

*Source*: Used with permission of G. Novakovic.

and in 2005 we launched Tesla.[13] We defined Tesla as 'an informal art and science discussion forum dealing with visionary ideas beyond the existing remits of art and science, that welcomes artists, scientists, theorists and curators and others active or interested in the field'.[14] The format of the meetings, held approximately monthly during the university year, is usually based around a presentation by an invited speaker, followed by an open discussion. Attendees include staff and students from UCL and other universities, in addition to a range of other interested individuals from London and beyond. All Tesla events are announced on the Computer Science department's official website and through a range of networks. We have also held some larger events, such as the well-received AHRC Methods Network Symposium 'Visions and Imagination: Advanced ICT in Art and Science', held in November 2007. The independent artist and documentary film-maker Amanda Egbe has now

13 The group is named after Nikola Tesla, the Serbian scientist, engineer and inventor, <http://www.britannica.com/EBchecked/topic/588597/Nikola-Tesla >.

14 The Tesla Art and Science Research Interest Group, <http://www.cs.ucl.ac.uk/research/tesla/>.

joined the core Tesla team. Her creativity, experience, enthusiasm and commitment to online documentation and archiving have brought a new momentum and a new mission to Tesla. We are now building an online video archive of all our events, generously supported by the Centre for e-Research at King's College and UCL.

## The past and the future

At the moment, I am once more exploring the original concept of a 'tuneable' artwork, of 'calibrating' an interactive installation in order to examine the changed and changing nature of contemporary perception and cognition. My interest in phenomenology first led me to neurophenomenology, then to the neuroscience of Gerald Edelman,[15] and eventually to neuroplasticity, the brain's amazing capacity for actively rewiring itself through its exposure to and interaction with the environment.[16] After the Belgrade exhibition, I realized that many of the techniques used in *Fugue* could have powerful neuroplastic consequences, and I am now thinking in terms of using them in a controlled way to produce known (and desirable) effects. In 2009 I wrote:

> … the dramatic shift in neuroscience brings with it a fascinating opportunity to explore and analyse the effects of electronic media through scientifically informed art, which could give rise to an entirely new art form: neuroplastic art. The concept of neuroplastic art opens a future for scientifically articulate artists and artistically articulate scientists to work closely together, with a full awareness of both the potential and the danger that emerges from the parallels between the nature of our nervous system and the characteristics of digital technology and electronic media. It may be possible to structure art works according to new scientific evidence, and to fuse scientific knowledge with imagination to exploit the nature of electronic media to create platforms for experiences that have never existed before. Bringing together the scientists' knowledge about the brain, and our knowledge of the properties of electronic media, we can envisage art works that will become tuneable complex instruments, serving both art and science. Only then will imagination and creativity transcend today's mere fascination with state-of-the-art technology, and use both technology and brain science as a means to express ideas. Perhaps this will even uncover new and benign ways of linking our brains with, and through, technology …[17]

15 G. Edelman, *Bright Air, Brilliant Fire: On the Matter of the Mind* (New York: Basic Books, 1993).

16 N. Doidge, *The Brain That Changes Itself* (New York: Viking, 2007).

17 G. Novakovic, 'Metropolis: An Extreme and Hostile Environment', in *MutaMorphosis: Challenging Arts and Sciences*, Conference Proceedings <http://mutamorphosis.wordpress.com/2009/01/15/gordana- novakovic/>.

Looking back at my work over the four cycles, I can see a major shift in the form of a transition from intuitive explorations to focused research; from artworks inspired by science to those informed by science. One of the major advantages of my residency is the privilege of being in contact with world-class scientists, something that is very difficult to do from outside academia. My contact with them, and my exposure to their work and to their working methods, has definitely coloured my thinking, and I am sure that this experience will continue to both form and inform my future work.

# Chapter 8
# Limited Edition – Unlimited Image: Can a Science/Art Fusion Move the Boundaries of Visual and Audio Interpretation?[1]

Elaine Shemilt

Since graduating from the Royal College of Art in 1979 I have had a career as an artist/printmaker and have taught and have run printmaking departments in art colleges. I have always been interested in how artistic reinterpretation can offer new insights into the analysis of scientific data. The following short history of printmaking may help explain how this may be done. Printmaking by mechanical means onto paper was adopted and developed in western Europe around 1400. It derived from earlier techniques of stamping or rubbing onto cloth, which were in common use in Europe by 1300.[2] Printmaking quickly developed as a means of imparting information and ideas, and for the motivation of piety and reflection, as can be seen in the 'Biblia Pauperum' or 'Bibles of the Poor' which are amongst the first examples of the practice of block printmaking. However, early processes such as woodcut, where images were incised into wooden blocks, were fairly crude. The very nature of the technique meant that the viewer was not provided with much more than an iconic or graphical representation of the object depicted. By the first quarter of the fifteenth century, more refined technologies, such as copper engraving, were adopted for printmaking, followed later in the same century by etching. These more sophisticated techniques allowed far greater subtlety of execution. The world was changing – as the media theorist Marshall McLuhan put it: 'The increasing precision and quantity of visual information transformed the print into a three-dimensional world of perspective and fixed point of view.'[3]

The next major printmaking breakthrough came at the end of the eighteenth century in 1796, when Aloys Senefelder (1771–1834), a Bavarian dramatist, developed the process now known as lithography. He was motivated by the fact

1 This chapter further develops a paper published in A. Bentkowska-Kafel, T. Cashen, H. Gardiner (eds), *Digital Visual Culture: Theory and Practice* (Bristol: Intellect, 2009).

2 A.M. Hin, *An Introduction to a History of Woodcut* (New York: Houghton Mifflin, 1935; New York: Dover Publications, 1963), pp. 64–94.

3 M. McLuhan, *Understanding Media* (London: Sphere, 1971), pp. 173–4.

that it was very expensive to make sufficient copies of his plays for his actors using available printing technology. Over a period of fifteen years, through a mixture of 'chemistry and physics, art, craft, skill and luck',[4] Senefelder developed the process which allowed him to create multiple copies of an original which was drawn upon a flat stone surface.[5] For a draughtsman, this is the probably the most natural of all the printmaking processes because it is most like drawing. The artist can work with pencil, ink or crayon, and incorporates the tonal variation and nuance of the mark at the time of drawing. Stone lithography became very popular toward the end of the nineteenth century and remained so into the twentieth century. As a result of this breakthrough there was a revolution in the dissemination of images including offset lithography, utilizing thin metal plates and high-speed printing machines, along with the photographic transfer of images.

It is important to recognize that the basis of artists' printmaking is still largely the traditional craft of woodcut, linoleum block, etching, stone lithography and screenprinting. However, at the beginning of this new century most artists use these techniques combined with digital-imaging processes to address new dimensions and contexts. Photomechanical techniques were the mark of twentieth-century printmaking. However, in the twenty-first century we live in an age of electronic and digital technology, new media, and new installation strategies. It has been possible to incorporate printed elements, for example screenprints and etching, into installation and public works. Currently artist/printmakers define themselves in many ways. There is a renewed interest in traditional printmaking techniques such as mezzotint, chine collé and photogravure while simultaneously we see rapid developments in the field of non-toxic printmaking technologies. Printmakers also now appropriate materials such as photo polymers and commercial silicon to develop new methods of printmaking in the experimentative tradition of Senefelder. Contemporary printmaking has incorporated ancient and modern techniques and relies upon an increasing precision and quantity of visual information to convey complex ideas and insights. In particular, digitization and advanced computing technologies have transformed the processes of screenprinting, lithography, and relief and intaglio printmaking. Simple mono-prints, multimedia prints and more complex lithography techniques are facilitated and accelerated by the use of computer technology. Techniques applied in the course of the emergence of digital art have generally sought to emulate the traditional processes used to create a printed image. The dramatic development of lithography in the pre-electronic age is reflected by the development of the new 'stone' of the latter half of the twentieth century and digital age: silicon. Digital technology, thanks to the opportunities it gives to combine art with electronics, algorithms and binary code, has reinvented print and brought with it the next revolution in image production, manipulation, dissemination and distribution – in short, contemporary printmaking.

4 M. Banister, *Practical Lithography Printmaking* (New York: Dover, 1972), p. 4.

5 P.B. Meggs, *A History of Graphic Design* (New Jersey: John Wiley and Sons, 1998), p. 146.

As a contemporary artist/printmaker, I am naturally interested in the visual transformation of printmaking through the use of digital technology. I am also interested in printmaking when it is used to illustrate someone else's ideas, and printmaking when it is used in collaboration with other disciplines in order to move the boundaries of interpretation. The latter includes my own definition of printmaking, which among other aspects, attempts to convey complex ideas and insights, incorporating scientific data, and progresses beyond the still print into digital animation.

This obviously involves collaboration with scientists. Historically, painters and sculptors have worked collaboratively with printmakers in order to make graphic reproductions of their work. The printmaker used his technical skill to create an image corresponding to the given brief, producing numbered editions of prints. In commercial terms, the numbering of prints safeguards their value. Professional artists' plates or blocks are usually cancelled after an edition has been completed. This convention adds the necessary 'aura' (in Walter Benjamin's famous use of the term) to make each single print a work of art. Benjamin's use of the term 'aura' came well before the technological ability or capacity to make an infinite, exact reproduction through digital means. Is it, perhaps, precisely because of the digital revolution that the ideal of the aura still seems to retain philosophical weight?[6]

Alongside the above method there are other ways in which collaborative printmaking has been undertaken. I believe that the techniques applied in the course of the emergence of digital art are not only an emulation or development of the traditional techniques used to create the printed image, but that, now more than ever, the mental and physical act of coordinating a germinal idea is a fundamental part of the role played by the printmaker as artist/collaborator/illustrator. This role does not, in my opinion, come as easily or naturally to artists working in media other than printmaking. I believe it has developed over time through the historical practice of collaborative printmaking.

This tradition of the 'limited edition' continues but contemporary printmaking neither needs to be defined nor confined by tradition or medium. The essence of contemporary printmaking lies in a process of empirical experimentation, discovery, analysis, resolution and critical reflection. The limitations placed upon the artist in this sense are therefore mostly technical. In order to illustrate the impact that successive technological breakthroughs have had on printmaking, it is perhaps important to note a few of the most important examples.

Examples of printmaking that illustrate an idea are numerous throughout the history of art (and science) and have had a bearing on technological innovation. One suite of prints particularly interests me: the series of illustrations which accompany Voltaire's novel *Candide*, created by the artist Paul Klee. It was my discovery of these etchings 30 years ago that introduced me to printmaking. The

6 W. Benjamin, *The Work of Art in the Age of Its Technological Reproducibility, and Other Writings on Media* (ed.), M.W. Jennings, B. Doherty and T.Y. Levin (Cambridge, MA: Harvard University Press, 2008).

26 illustrations to *Candide* (completed in 1911) were the first great landmark in Paul Klee's graphic output. These illustrations were the conclusion to Klee's search for 'progress in the line'[7] before he developed the confidence to pursue a career as a painter. By 1911, Klee had developed a graphic language that could serve representational ends without conflicting with his abstract functions or with his expressive needs. Klee found the task of illustrating *Candide* to be a crucial stepping-stone in the direction of his later work. *Candide* was the satirical vehicle for Voltaire's outrage, exasperation and, at heart, love. Klee, from the first, considered this book to hold tremendous potential for illustration. He did not aim to produce a picture which gave form to an idea, in that it represented a theme or illustrated a story; his intention was that the significance of his work would be created and determined by the pictorial composition itself. This was traditionally a significant factor in the role of a printmaker.

The etchings by Klee for *Candide* are relevant for the following reasons. In art the content of a picture or the storyline of a book counts for little or nothing if the style, that is the manner in which the subject is drawn, painted or described, is not original to the artist. It can be nurtured by study, but it cannot be contrived, artificial or borrowed from someone else. The action of printing is metaphorical, that is to say the repeated printings are an allegory of the first. They are identical in all relevant aspects to the original notation. However, for Paul Klee, the illustrations to *Candide* were the key to his 'progress in the line'. *Candide* did not provide Klee with any answers within his drawing, but it did help him to deal with other aspects of his artistic aim. Success in illustrating Voltaire's *Candide* led Klee to success as a painter. The etchings to *Candide* are a one-sided collaboration, for Voltaire had been dead for almost 150 years, but they do give some indication of the aim Klee pursued in graphic art.

The role of the collaborative printmaker has always been multifaceted and unclear, and has seldom been given due recognition. However, in the twenty-first century, printmaking encompasses a variety of processes and it offers visual flexibility. The printed image is everywhere in contemporary culture and is frequently mediated by computer technology. The desktop printer is an obvious example, but as important are the processes of layering colours and registration, mark making and separation techniques. Once the digital proxies developed, then the potential from the artist's palette, as it were, became limitless.

I would like now to return to the nature of artistic collaboration and the impact of digital technologies upon the nature of collaboration. I also wish to consider the distinctive forms of art practice that have developed within research environments. Usually, this research takes place in the safe confines of higher education institutions. For many years, examples of research applications to funding bodies had already been set by scientists. Many artist academics have learned from their science colleagues and have developed their ability to make successful bids by

7 P. Klee, *The Diaries of Paul Klee 1898–1918* (ed.) Felix Klee (Berkeley, CA: University of California Press, 1964), p. 260.

using science applications as templates. Added to this, advanced ICT methods now enable collaborations of a sort previously unimaginable. The term 'research methodology' has become familiar and is now used as frequently by practising artists in higher education institutions as by any other academic.

This leads to my own experience. A number of years ago I began to collaborate with three scientists. Working at the Scottish Crop Research Institute (SCRI) in Dundee, Scotland's leading centre for plant research, the scientists were investigating a bacterial pathogen, *Erwinia carotovora* (*Eca*, subsp. *atroseptica*), that causes a devastating disease in potatoes. Interestingly, *Eca* belongs to the same group of bacteria as the deadly human pathogens *Escherichia coli* (*E.coli*), *Salmonella* and *Yersinia* (bubonic plague pathogen). Enterobacterial pathogens, whether causing disease in humans, animals or plants, share over 75 per cent of their DNA. Why then should the potato *Eca* organism, so close in its makeup to the human pathogens, be harmless to us and yet be so dangerous to plants? Presumably the answer lies in the remaining 25 per cent unique complement of genes. The aim of the scientists was to find the genes that are present in *Eca* but absent in the animal pathogens, and accordingly determine the roles of these genes in disease.

Dr Ian Toth is an expert on bacterial pathogenesis, molecular approaches to host/pathogen interaction, genome sequencing and functional genomics. Dr Leighton Pritchard is a scientist working at the interface of biology and computer science, concentrating on how microorganisms cause disease in plants. Pritchard and Toth set out to determine the identity and position of every gene on the chromosome of *Eca*. The *Erwinia* sequencing project began in 2004 and they found themselves in a transatlantic race to sequence the organism. They succeeded in becoming the first in the world to hold the genetic 'blueprint' for *Eca*. The analytical tool that these scientists developed, 'Genome Diagram', has been adopted by an increasing number of genomics labs internationally (e.g. the Sanger Institute in Cambridge, England, and the universities of Minnesota and Madison in the United States). However, the Scottish crop researchers can have difficulty in communicating their complex scientific discoveries to non-experts. When published, their articles are usually targeted towards a particular audience such as growers, but it is through publication that the scientist's position and value in society is assessed. There has been a recent shift in the focus of the Scottish Executive Environment and Rural Affairs Department towards more publicly accessible topics such as climate change, and of course Scottish crop researchers also benefit if they can show that money and effort are being spent with the needs of the general public in mind.

Dr Michel Perombelon, a senior retired scientist at the SCRI, approached me with a view to collaboration. Dr Perombelon had held discussions with Dr Ian Toth and Dr Leighton Pritchard, who hoped that artistic expression and communication might help the wider community understand a complex scientific discovery and at the same time generally raise awareness of their research. At the time, this was part of a radical move towards the dissemination of science to a larger audience, both locally and nationally. I was inspired by the images that the scientists had created to represent the genetic data. I could see that the thousands of tiny bits of

DNA, represented in diminishing circles and lines across the computer screen, lent themselves to visual exploration. The beauty created by an evolutionary process is particular. I do not, of course, have the same view of the science as the scientists, but as an artist/printmaker I am used to artistic freedom and have a tendency to think in pattern and subtleties of colour, much in the same way that anyone can add to their understanding of a sculpture if it is looked at from a variety of viewpoints. During this time the scientists and I spent a considerable time getting to know each other and our respective research specializations.

My exploration began with a series of screenprints in which I paid very little regard to the meaning of the scientific data, although I was careful to retain the correct order of the genetic sequencing. I transposed the pattern and concentrated on developing my own shades of colour and tonal variation. [Plate 8.1]

Looking at my screenprints, Dr Pritchard noticed a feature that my image had highlighted. This previously-overlooked feature allowed him to recognize the occurrence of new pathogenicity determinants. As he observed

> the abstraction of the physical, sometimes invisible world into an easily visualized and understandable form, free of metaphor and exaggeration is the stock-in-trade of the scientist. Without formal training in alternative modes of visualization, some insights that could be readily revealed may escape us. By the artist's use of iridescent ink overlays, the collaboration has already introduced us to a mode of presentation that was previously unknown to us. Patterns of gene conservation were revealed to us that would have gone unnoticed in our original diagrams. As the process of abstraction influences the mode of visualization, the form of visualization may affect the future process of abstraction. We expected that greater insight into our own processes of abstraction and analysis of the data itself would flow from this collaboration.[8]

It was in this way that our collaboration quickly reached beyond the initial objective to raise public awareness, into deeper research issues on the possible role of the artist in the visualization of complex data, and the subsequent impact upon scientific understanding and insights.

The question arose for us as to whether a science–art fusion could move the boundaries of visual and audio interpretation to any significant degree.

Dr Pritchard works at the interface between biology and computing. His first thoughts when this project was suggested concerned the aesthetic value inherently present in scientific information, even in the absence of a context. The presentation of scientific information has a deserved reputation for being literal and representational, with a minimum of embellishment and extrapolation. This is often required for the clear and precise dissemination of accurate information. The guiding theme in preparing scientific figures for publication is often that they should be interpretable without reference to the main text. Why would a

8 Email communication, Dr Leighton Pritchard, November 2005.

scientist approach an artist/printmaker with a view to collaboration? At this stage the scientists were working with computer images that were perfectly adequate for publication. The explanation as I see it does not elude art history. With the advent of digital technology, contemporary printmaking incorporates ancient and modern techniques and allows for an increasing precision and quantity of visual information to convey complex ideas and insights.

As mentioned previously, Dr Ian Toth is an expert on bacterial pathogenesis, molecular approaches to host/pathogen interaction, genome sequencing and functional genomics, while Dr Leighton Pritchard works at the interface of biology and computer science, concentrating on how microorganisms cause disease on plants. The software, 'Genome Diagram' enables simultaneous visualization of billions of gene comparisons of hundreds of fully-sequenced bacterial genomes, including those of animal and plant pathogens. The acquisition of foreign DNA by pathogens, potentially representing novel mechanisms that are involved in disease, are represented by clearly defined white 'spokes' radiating from the centre of the image. [Figures 8.1, 8.2 and 8.3]

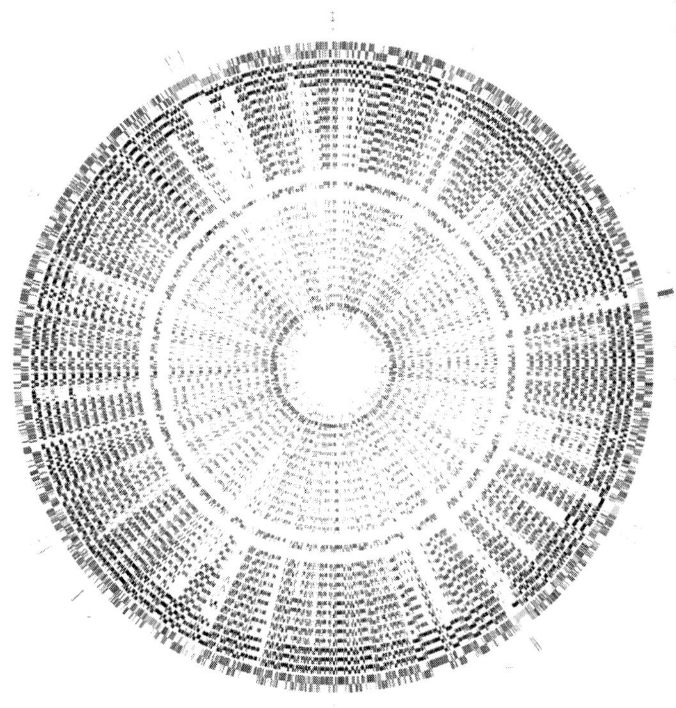

**Figure 8.1** ***E. Coli 1*, 2004**

*Source*: Used with permission of Leighton Pritchard, SCRI.

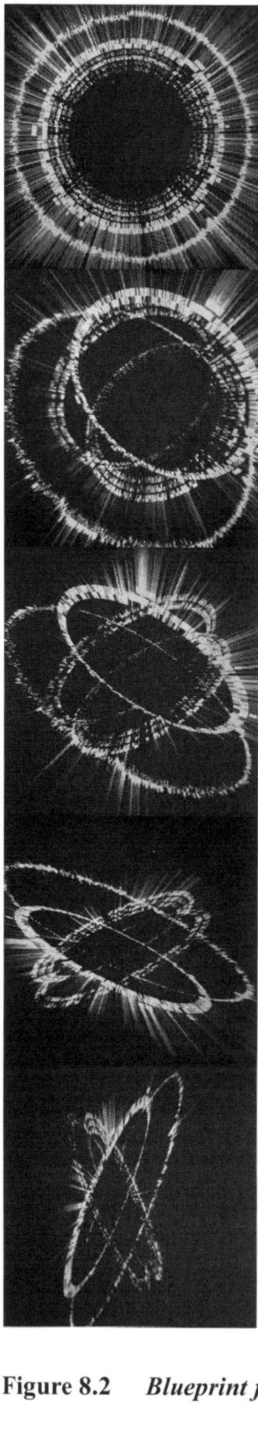

In this way it is also possible to trace the evolution of this gene acquisition (and loss) over millions of years. At root, DNA transfer is the single most significant source for the outward differences between diseases caused by closely related bacteria. Extraordinary as it seems, the acquisition of foreign DNA may culminate in a microbe changing into either a human or plant pathogen and the point at which this occurs is a 'tipping point' in that microbe's evolution. This foreign DNA in turn leads to novel biological traits being introduced into the microbe and incorporated into existing regulatory circuits such as quorum sensing. As pathogen populations grow in their host, they produce a regulatory hormone that gradually increases in concentration. At a critical (or quorate) population, the concentration of that regulator hormone becomes sufficient to trigger a series of events essential to symptom development and disease initiation. The point at which this trigger occurs and true disease begins is the 'tipping point'. Our collaboration attempted to emphasize a parallel between the artistic event of visualizing the biological data and the biology itself.

The images produced by Genome Diagram, even in their scientific context, are fairly abstract. Most of the processes and entities with which modern microbiology is concerned, are invisible to the naked eye. Aspects of genomics are similarly invisible. Each genome is the result of four billion or so years of evolution. It is a huge concept to grasp even when it is made visual in a diagram. In the first series of prints I consciously put out of my mind the scientific information relating to the Genome Diagram. I removed the letters and codings so that I could treat it as an artistic project. It was a scientific image stripped of its contextualizing information. In other words the image, a circular map of genes and their relationship to other bacteria, represented something essentially invisible that could only be 'seen' in an abstract representation. I began by concentrating on subtleties of colour and tonal variation. Then I focused on the precision and

**Figure 8.2** ***Blueprint for Bacterial Life*, 2006. Screenprint**

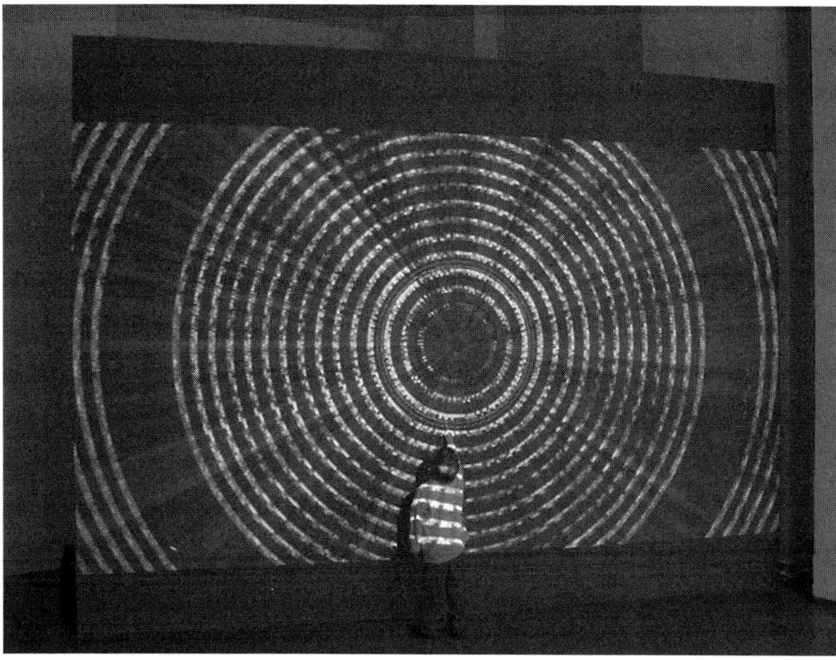

**Figure 8.3** ***Blueprint for Bacterial Life*****, 2006. Still from animation with child viewer**

*Source*: Photograph by Ian Toth.

quantity of visual information and created a series of etchings, screenprints and animations. With the screenprints I used a very subtle range of silvery blues and greys and worked with inks known to printmakers as 'interference inks'. These are metallic and impart a slight three-dimensional quality. [Figure 8.4]

My approach was to simplify the diagram into a tonal variation. In so doing I re-contextualized the data in such a way that the prints revealed information to the scientists which they had previously overlooked in their systematic and empirical approach to the data. Rather than simply identifying genes unique to a pathogen, the screenprints revealed the presence of other genes present in all of the bacteria, possibly representing genes essential to all forms of bacteria.

By taking this scientific visualization tool outside the fields of biology and medicine and placing it into the context of interdisciplinary art, both artist and scientists entered into new territory. Inspired by this, we began exploring the dynamic nature of biological systems using both visual and sound disciplines (and their associated media) and went beyond obvious interpretative frameworks. Our goal was to ensure that the relationship of the artwork to the data was reflected and maintained not merely as content but also as elements and structural process. I began work with Daniel Hill, an animator who had recently graduated from the MA

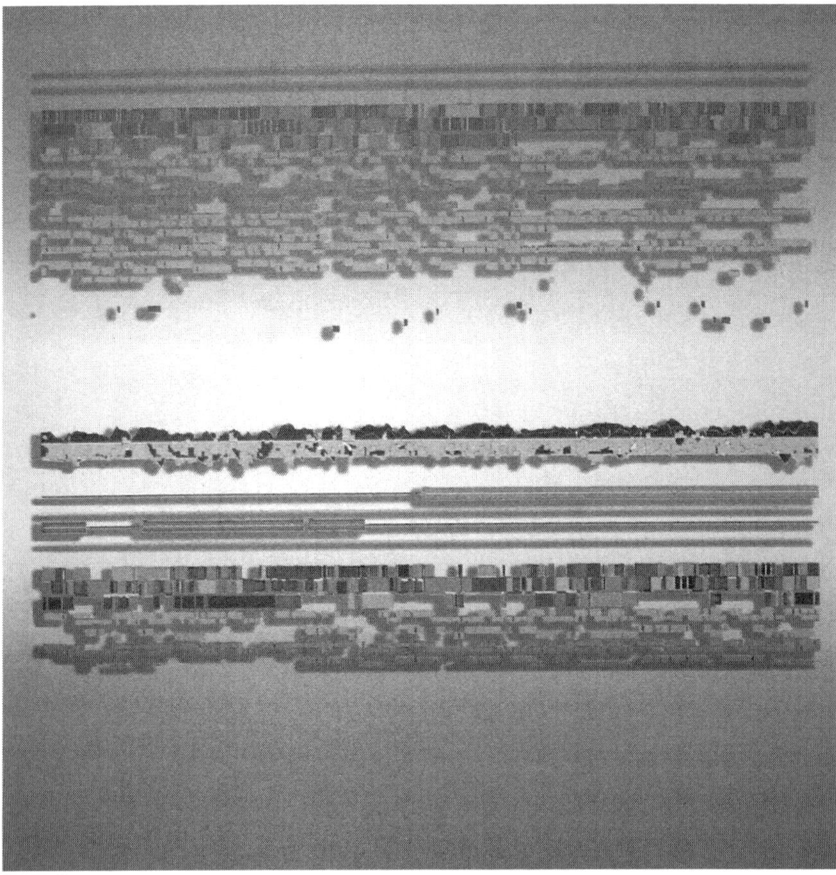

**Figure 8.4 *Linear Blueprint*, 2005. Screenprint**

programme in Animation and Visualization at the University of Dundee. The aim was to progress to the more complex systems arising from the emergent properties of the sequences, but initially we concentrated on a simpler representation, that of of sequential gene expression in disease development. The animation consisted of the gradual construction of the circular DNA diagram from the core outwards and finishing with the *Erwinia* DNA, but with gaps representing the acquired DNA in the circle. This was then followed by the deconstruction of the core, with bits of DNA moving randomly towards the periphery to fill in the gaps in the *Erwinia* DNA circle, ending with what we know of the *Erwinia* genome. In another short animation the *Erwinia* DNA ring coils up in the form of a hoop – like super-coiled DNA – before unwinding to restart the cycle. If these processes could be presented in continuously repeating cycles involving different DNA on each revolution, we envisaged that this would carry the notion of evolution continuity. [Plate 8.2]

We also began working towards an interactive sound and image environment. The idea was that either by accident or intent, spectators, by their presence and by certain movements, would be able to select or direct which bit of DNA could be incorporated into the environment in order to introduce new properties. The effect would be to weave the participant in and out of an illusory immersive space while triggering sensations of colour, light and sound.

From some perspectives the DNA image resembles a score of music. Dr Pritchard trained originally as a chemist and has a versatile view of biological information that is both physical and chemical. By using a series of mathematical notations he translated the different amino acid letters into sequences of musical notes. As he puts it:

> Aside from the biological and physical meaning of this letter 'A' it is not even represented as a letter inside the computer. When my finger hits the 'A' key on the keyboard it initiates a series of electrical pulses. The computer then interprets these pulses as a binary number. When we need to 'write' the character to the screen, a different series of electrical pulses are used. These represent not the letter itself, but an image – patches of light and dark on a larger canvas. The use of different font types will result in different patterns, and so different pulses, but still the same recognisable symbol. These representations are at once inexact but precise.[9]

We asked a student of classical music, Genevieve Murphy, to work on Leighton's musical note sequences. By developing the scale, tonality and starting octave of the melody, and the intervals for each base transition, she translated it into the auditory sphere and created a series of musical compositions which were performed using tuned percussion. At the same time we gave the data to David Cunningham, an artist who specializes in the manipulation of sound by electronic and acoustic process. He places a particular emphasis on the integrity of the materials, their innate structure and context. This emphasis on process is a key element in this project, an approach that can creatively maintain the precision of the source data.

The challenge and motivation for developing the installation cleared the way for a process of experimentation, and techniques involving time and space. It became possible to introduce sound and spatial aspects through open, interrogative and responsive modes of thinking. To begin with we dealt with the linear data of the genome sequence, creating images, animations and simple sound, based upon the data translated through MIDI (Musical Instrument Digital Interface).

My enthusiasm, as a printmaker, lies in creating a diversity of ideas and attitudes, which printmaking techniques can communicate. The layering process of printmaking allows for the aesthetic manipulation of the image in the form of animation. By taking the data from Genome Diagram and separating the information into layers, I created the first series of screenprints by means of very careful

9 Email communication, Dr Leighton Pritchard, November 2005.

registration. Screenprinting differs from other printmaking processes in which the print is taken by the impression of one surface upon another. The principle of screenprinting is that the ink passes through an intermediate surface, i.e. the screen mesh. Images are created by blocking out areas of the screen mesh and leaving the image area of the screen open for the ink to pass through. Originally this was done with paper or stencil film, but now printmakers generally use a light-sensitive liquid that fills the screen mesh, and the image is processed digitally or photographically. Screenprinting still has a direct association with commercial and industrial printing processes and had developed alongside photomechanical lithography as a means of mass image production. However, a problem with screenprinting was that in the registration of prints that required more than a few colours, the half-tone screen, or dot-matrix stencils, always produced moiré patterns if one of them was printed over another. The switch from a photomechanical process to digitization solved this problem overnight. Immediately artist/printmakers found a freedom of expression unhampered by the problems of registration and moiré effect. There was nothing to get in the way of multi-layered, multi-coloured screenprints with exquisitely intricate detail. Genome Diagram would not have been possible without the advent of digital technology. Prior to this, it would have been impossible to render the intricacy of the detail or to retain the vast amount of information found in the diagram. Printmakers today have access to a far greater range of methods than ever before, and inevitably we have come to realize that certain ideas are best developed through a combination of several techniques. Screen inks cover well and although the choice of technique obviously depends upon the effect that is required, I have found that some of the best combinations are created by screenprinting on top of lithography or intaglio. Metallic and fluorescent inks, most effectively used through the screenprinting process, can be printed on to any flat dry surface. The layering processes used in translating the output from Genome Diagram into screenprints provided the basis for the layering processes required to make the animations. Digital technologies provide flexibility and also allow for a convergence of media that previously suffered cumbersome, awkward relationships. Within the digital domain it has been possible to develop the concept of Genome Diagram into a multimedia interactive installation, with animations and music based on the genetic plasticity and evolution of bacterial pathogens.

The unifying thread of our collaborative art–science research is that by decontextualizing scientific data, we obtain a complementary viewpoint to that of the scientific interpretation. Both disciplines thrive on lateral thinking and observation. As well as refining our mechanisms for creative development, our collaboration aims to enhance scientific visualization of complex data and promulgate scientific understanding and insights. Common to both artist and scientist is the use of advanced visualization tools and the principles of New Media, which have been defined as 'numerical representation; modularity; automation; variability; and cultural transcoding'.

Research development continues to involve production, analysis of visualizations in print, digital imaging, 2D and 3D (HiDefinition) animation and sound. By using animation to create time-lapse video clips the intention is to create new dimensions for the expression and interpretation of the data. The rapid development of computer technology and computer graphics has enabled advanced visualization techniques; an essential part of the huge data-generating potential of genomic technologies. Both scientists and artists are exploiting the latest technologies. This project enables us to share and resolve the problems that surround the current uses of visual and audiovisual techniques when approached from different perspectives. I use current digital reproduction techniques in my prints in order to address the scientists' research. The scientists are reliant on some of the most cutting-edge computer technology to create the most advanced comparative genomics visualization tool worldwide.

The development of digital printmaking has enabled a reinvention of the artist's language. What does the aesthetic manipulation of the image look like in the twenty-first century? Artists will always continue to strive to communicate on social and psychic levels. With the advent of digital technology printmakers are no longer the ancillaries of the art world. They have as much scope as any artist has ever had. They can create continuous experiences of movement, simulate human consciousness, and, with the use of current technology they can even create artworks that are blueprints for bacterial life.

Chapter 9

# Telematic Practice and Research Discourses: Three Practice-based Research Project Case Studies

Paul Sermon

## Introduction

This chapter focuses on the production, documentation and preservation of the author's telematic, practice-based research in the interactive media arts. It reflects a timely practice review with significant implications for the future of exhibiting and archiving the broad range of creative arts in this field. These fundamental research questions also have relevance across a number of practice-based research fields including performance arts and the ephemeral nature of open-system interactive artworks. The objective of this chapter is to propose research methods that will approach the question of how to document and archive appropriately this transient creative practice that is so often reliant on its immediate cultural and historic context.

Since the early 1990s my artistic practice has identified and questioned the notions of embodiment and disembodiment in relation to the interacting performer in telematic and telepresent art installations. At what point is the performer embodying the virtual performer in front of them? Have they therefore become disembodied by doing so? A number of interactive telematic artworks will be looked at in detail in this chapter. These case studies range from Kit Galloway and Sherrie Rabinowitz's seminal work *Hole-in-Space* to my own telepresent experiments with *Telematic Dreaming* and include the current emerging creative/critical discourse in Second Life, the networked virtual/social environment, that polarizes fundamental existential questions concerning identity, the self, the ego and the (dis)embodied avatar.

The preservation and documentation of this work is extremely problematic when we consider the innate issues of (dis)embodiment in relation to presence and intimacy, as experienced and performed in telematic and virtual environments. How can it become possible to reencounter a performance of dispersed and expanded bodies, multiple and interconnected identities, spectral representations and auras; in short, hybrid bodies (selves) made of flesh and digital technologies, and the intimate connections between them?

## Telematic practice

My work in the field of telematic arts explores the emergence of a user-determined narrative by bringing remote participants together in a shared telepresent environment. Through the use of live chroma-keying and videoconferencing technology, two public rooms, or installations, and their audiences are joined in a virtual duplicate that turns into a mutual, visual space of activity. Linked via an Internet videoconference connection, this form of immersive interactive exchange can be established between two locations almost anywhere in the world.

The audiences form an integral part of these telematic experiments, which simply would not function without their presence and participation. Initially, the viewers seem to enter a passive space, but they are instantly thrown into the performer role by discovering their own body-double in communication with another physically remote user on video monitors in front of them. They usually adapt to the situation quickly and start controlling and choreographing their human avatar. The installation, which is set up in the form of an open accessible platform, also offers a second choice of engagement: the passive mode of just observing the public action, which often appears to be a well rehearsed piece of drama confidently played out by actors. Compelling to watch, it can be a complex issue to discover that the performers are also part of the audience and are merely engaging in a role. The entire installation space then represents two dynamic dramatic functions: the players, controllers or puppeteers of their own avatar, absorbed by the performing role; and the off-camera members of the audience, who are themselves awaiting the next available slot on the telematic stage, soon to share this split dynamic. However, the episodes that unfold are not only determined by the participants, but by the given dramatic context. As an artist I am both designer of the environment and therefore 'director' of the narrative, which I determine through the social and political milieu that I choose to play out in these telepresent encounters.

## *Headroom*: A space between presence and absence (2006)

This case study represents the first theoretical account of *Headroom*,[1] a site-specific interactive art installation produced in Taipei by myself, as the successful recipient of the 2006 Taiwan Visiting Arts Fellowship. This residency programme was a joint initiative between Visiting Arts, the Council for Cultural Affairs in Taiwan, the British Council (Taipei) and the Arts Council, England. The development of this interactive art installation was extensively documented as part of the AHRC Performing-Presence project led by Professor Nick Kaye

1 Artist's website and documentation <http://www.paulsermon.org/headroom/>. (All URLs current at the time of writing.)

from Exeter University in partnership with Stanford University.[2] *Headroom* was exhibited at Xinyi Assembly Hall, Taipei in April 2006.

*Headroom* is a juxtaposition of the artist's ethnographic research experiences in Taipei, between the ways people 'live' and the ways that they 'escape' the city, as an analogy between the social networking telepresent aspirations of the 'headroom' (Internet) space and the solitude of the 'bedroom' (private) space. Referencing Roy Ascott's essay, 'Is There Love in the Telematic Embrace?' (1990),[3] and reminiscent of Nam June Paik's early TV-Buddha installations,[4] *Headroom* is a reflection of the self within the telepresent space, as both the viewer and the performer. The television 'screen' is transformed into a stage or portal between the causes and effects that simultaneously take place in the minds of the solitary viewers. The installation overtly intertwines private and public space, and the sense of the 'inside' and 'outside' of the installation's 'place'.[5] It is partly in this breaking down of oppositions that the participants' sense of the 'presence' of their co-performers is amplified. In this aspect, *Headroom* radically extends a disruption of oppositions in which video art/installation and site-specific work has frequently operated. The co-performers discover themselves acting out a series of intertwinings of public/private, inside/outside. The installation itself and its title emphasize the intimate nature of this overlaying of spaces – the aspect of fantasy or dream – while the public nature of the installation sanctions or appears to give permission or consent to this closeness. In this context, co-performers discover themselves 'coming closer' in a paradoxical distribution of presence – an intimacy produced by a telepresent distance. Here, visitors discover themselves occupying and acting out their co-performer's private space, while seeing their own private space acted out by their telepresent partner. The spatial rules of public interaction are breached, producing an intimacy, a particular and shocking closeness, and a dialectic between the explicit sense of being here (in the bedroom, for example) and being there (acting out the space of the other), while seeing and responding to their co-performer's mirrored reaction. [Plate 9.1]

Located in the east of Taipei city in the shadow of the 101 Tower and Taipei's World Trade Centre is a Taiwanese war veterans' housing complex built toward the end of the 1940s. This site has been renovated and converted into a museum and exhibition space. It sits on some of the most commercially sought-after space in the city, but because of its historical importance to the liberation of Taiwan it remains a listed building. The back-to-back terraced streets have been knocked through into entire buildings, creating three large exhibition halls that retain the

---

2 AHRC Performing-Presence project, <http://presence.stanford.edu:3455/Collaboratory/500>.

3 R. Ascott, *Telematic Embrace* (Berkeley: University of California Press, 2003), pp. 232–46.

4 N.J. Paik's TV-Buddha, <http://www.medienkunstnetz.de/works/tv-buddha/>.

5 N. Kaye, *Site-Specific Art: Performance, Place, Documentation* (London: Routledge, 2000).

original appearance of the houses on the outside. The spaces that interested me most were the small facade rooms created by the larger space conversion, which have been separated from the gallery space by interior glass walls and are only accessible from existing external front doors. The two facade rooms I used for the installation were identical in size and were used to house a connected telepresent installation where the audience participants in the separate facade rooms were unable to see each other. However, this allowed the audience inside the gallery to observe both participants in the space through the glass walls. The rooms were only about 2m by 3.5m wide, and 2.5m high. The original houses were longer, but no wider and the original inhabitants often halved the height of the rooms to create separate sleeping and living areas. This two-level use of the space interested me, and also reminded me of the outside of the space, with the 101 Tower in stark contrast to the little houses huddled around its base. This paradox can be seen in much of Taipei's culture, from very basic noodle bars and soup kitchens between Karaoke TV clubs, 7/11 convenience stores and high-rise office blocks to countless temples devoted to countless incarnations of the Buddha.

The project functioned by combining the two identical room installations within the same video image via simple videoconference techniques. The system worked as follows: The two rooms both had false ceilings lowered to a level of approximately 1.5m, which left a cavity space above each room of approximately 1m high and which forced the gallery visitors to bend down when entering the spaces. However, there was one location in each room where the viewer was able to stand up straight and put his or her head and hands through a hole in the false ceiling and into the cavity space above. Although each room shared identical dimensions, each had a strikingly different appearance. One of the rooms contained drab, used furniture in the lower part and had a very lived-in appearance. The cavity space above it was brightly decorated and had the appearance of a personal shrine or Karaoke bar. It had red curtains and a large video screen at one end. The other room by contrast, was empty in the lower section and very bright in the cavity above, including illuminated blue walls and another large video screen. A video camera in each space recorded a live image of the head and hands of each participant and fed this recording directly to a video chroma-key mixer. The background in the profile head-shot recorded against the bright blue walls was extracted by the video mixer and replaced with the other live profile head shot – placing the two heads opposite each other within the same live video image. [Figures 9.1, 9.2 and 9.3]

The red room represented a very theatrical, illusionary space. The blue room, by contrast, appeared to be a more functional backstage space. However, from the outside point of view there was not so much a front and backstage division as a juxtaposition of two entirely separate spaces, which, due to their sheer proximity, were meant to have something in common and yet, somehow, never became a telepresent synthesis. For Gabriella Giannachi there is a postmodern dialectic here,[6] expressed visually in the impossibility of the two spaces becoming one: the

6 G. Giannachi, <http://presence.stanford.edu:3455/Collaboratory/500>.

**Figure 9.1** *HEADROOM*, **2006. Video still**

**Figure 9.2** *HEADROOM*, **2006. Video still**

**Figure 9.3** ***Headroom*****, 2006. Video still**

external viewer, standing in front of the two spaces, actually sees 'nothing' but the real, whereas to see the telepresent space you actually have to be willing to be within it.

### *Liberate your Avatar* (2007)

Since May 2007 my practice and research has undergone what might appear to be a paradigm shift, focusing on the creative possibilities of the online multi-user virtual environment of Second Life. Whilst this represents a major departure from my established telematic projects, there are significant parallels between the earlier telematic video experiments and the presence and absence experiments currently being developed in Second Life. Together, these aspects of telepresence and the merger of 'first life' and Second Life aim to question fundamental assumptions of the Second Life phenomenon.

The aim of this project is to investigate critically how online participants in three-dimensional worlds, Second Life in particular, interact socially within innovative creative environments, appropriate these cultural experiences as part of their everyday lives, and question what is 'real' in this relationship. The project brings together ethnographic and creative practice-based methods that identify and develop original and innovative interactive applications, interface design and imperative cultural and sociological knowledge that will help shape and define the emerging online society and 'metaverse'[7] of Second Life, significantly contributing to the quality of both 'first life' and Second Life.

In Second Life you create an avatar that lives out an online existence. There are no set objectives; you can buy property, clothing, accessories, furnish your home, modify your identity, and interact with other users. This online community has grown to eleven million residents since launching in 2000, generating a thriving economy. However, whilst the virtual shopping malls, nightclubs, bars and beaches often reach their user capacity, there is a noticeable lack of creative and sociological modes of attraction and consequently the growing media attention around Second Life warns that this expanding community has become ambivalent and numbed by its virtual consumption and that there is an increasing need to identify new forms of interaction, creativity, cultural production and sociability.

However, when the Front National, the far-right French political party of Jean-Marie Le Pen, opened its Second Life headquarters in January 2007, the Second Life residents reacted in a way that would suggest they are far from complacent avatars wandering around a virtual landscape and that they possess a far greater degree of social conscience than the consumer aesthetics of Second Life suggest. Through prolonged mass virtual protest the centre was razed to the ground in the space of a week and has not returned since. The reaction to the Le Pen Second Life office begs the question: is Second Life a platform for potential social and cultural change? Is there a hidden desire and ambition to interact and engage with this online community at an intellectual and creative level that transcends the collective 'I shop therefore I am' apparentness of its community?[8] Moreover, does Second Life influence 'first lives'? If so, could our 'first life' existence start to reflect our Second Life conscience as this community continues to grow and develop into the future? As the landmass and population of Second Life expands at an ever-increasing rate, it is clear that essential research into the intersection and interplay between 'first life' and Second Life, and both new and old patterns of consumption, cultural production and sociability is urgently needed.

This second case study focuses on some of my most recent Second Life experiments entitled *Liberate your Avatar*, an interactive, public video art performance incorporating Second Life users in a real-life environment, as shown in Figure 9.2.[9] Located in All Saints Gardens, Oxford Road, Manchester, for the Urban Screens Festival, 12 October 2007 from 5.00 pm to 6.00 pm, this installation merged the realities of 'All Saints Gardens' on Oxford Road with its online three-dimensional counterpart in Second Life, and for the first time allowed 'first life' visitors and Second Life avatars to coexist and share the same park bench in a live

7 The term 'metaverse' comes from Neal Stephenson's 1992 classic science fiction novel *Snow Crash*, and is now widely used to describe the vision behind current work on fully immersive 3D virtual spaces.

8 The phrase 'I shop therefore I am' was used by artist Barbara Kruger in 1998 as a pun on consumerism and René Descartes' statement 'I think therefore I am.'

9 Artist's website and documentation, <http://www.paulsermon.org/liberate/>.

interactive public video installation. By entering into this feedback loop through a portal between these two parallel worlds this event exposed the identity paradox in Second Life. [Plate 9.2]

This unique project, commissioned by Let's Go Global in Manchester, brought together earlier practice-based telepresence research projects and current experiments and experiences in the online three-dimensional world of Second Life. The installation investigated the notion of demonstration and how it has been transposed from the real into the virtual environment. *Liberate your Avatar* exposed the history of All Saints Gardens; relocating Mancunian Suffragette Emmeline Pankhurst as an avatar within Second Life, where she remained locked to the railings of the park, just as she did a hundred years ago, reminding us of the continual need to evaluate our role in this new online digital society. *Liberate Your Avatar* examined this new crisis whilst drawing upon the history of the site, creating a rich, provoking and entirely innovative interactive experience.

The installation consisted of three specific spaces, two of which were located in the virtual world of Second Life while the third was physically located in All Saints Gardens on Oxford Road, in Manchester. The two virtual environments included a blue box studio and a three-dimensional replica of All Saints Gardens. They were located adjacent to each other, allowing the Second Life avatars to move freely between the two spaces. When an avatar entered the blue box space his/her image became chroma-keyed with a live video image from All Saints Gardens. This combined live video image of the avatar in the 'real' All Saints Gardens was then streamed back onto the Internet and presented on a virtual screen in both Second Life spaces. An image of the Second Life version of All Saints Gardens, with its virtual 'big screen' was then presented on the public video screen in the 'real' All Saints Gardens.

*Liberate your Avatar* brought together theoretical and practical methods from the field to address this identity crisis in 'first life' and Second Life. Although online communities have been studied in-depth for some time now, the focus here will be upon an ethnographic, multidisciplinary and practice-based discussion in order to paint a richer picture of future experiences.

In this respect, the project uncovered more question than answers, principally concerning identity and self. The ontological questions of virtual reality and identity, online or offline, have been at the centre of the contemporary media arts and science debate for the past three decades. *Liberate your Avatar* points at the social, political and cultural significance of Second Life by questioning the emerging relationship between 'first life' and Second Life as a platform for potential social and cultural change. Through this discourse the project questioned whether Second Life is a reflection of 'first life' or if 'first life' is actually a reflection of Second Life? By consciously deciding to refer to this mirrored image as 'first' life rather than 'real' life, this central question polarized the paradox in Second Life, considering Lacan's proposition that the 'self' (or ego) is a formulation of our own body image reflected in the mirror 'stage'.[10] However, there are no mirrors in Second Life, which raises the fundamental question of whether it is possible to formulate our second self (or alter ego) in Second Life at all. Or is the computer screen itself the very mirror we are looking at?

## *Hidden Voices: Memoryscape* (2006)

The final case study project, *Hidden Voices: Memoryscape*,[11] was commissioned by the Taipei City Department of Cultural Affairs for the fourth 'City on the Move' art festival, entitled 'From Encounter to Encounter– Expounding the Playground'. This took place at the Children's Recreation Centre, Taipei, Taiwan, in November 2006. *Hidden Voices: Memoryscape* invited visitors to enter the amusement park and, guided by PDAs and maps, to search for stories taking place amidst the physical terrain – for example, unusual past experiences that people had at the amusement park when they were children: 'a strawberry ice cream dripping on an orange skirt, a lost shoe, falling over and grazing a knee or how the space appeared then …'.[12] Stories and incidental experiences allow adults to reinterpret this place, which is the 'territory of children', while memories in synch with the archetypal concept of the venue induce the expansion of the subconscious, constructing an aesthetic of imagined memories in relation to the venue. Thus, the augmentation of individual memories is transformed into collective memory. In addition to the augmented mediascape, I presented a series of video-projected images in the tunnel of the miniature train ride. These video sequences referred to a momentary transition between the past and present experience of the amusement park and thus further assisted in augmenting the participants' journey around the environment. [Plate 9.3]

This project commenced by interviewing parents and visitors at the adventure playground over a one-week period and recording two-to-five-minute episodes about their own childhood experiences and memories of the adventure playground – intimate personal stories and strange and unusual memories about incidental experiences. In order to create this dynamic audio and video narrative, the work was partly constructed/dramatized and partly real-life stories/interviews.

This layering of augmented memories over the actual experience of visiting the adventure playground today was further assisted by providing visitors with a map that guided them through the locations and the stories attached to them. Whilst further conceptual information was provided in this guide, other discrete and unusual sounds and visuals that the user stumbled across were included, providing an abstract story or chain of events that brought the piece together

10 J. Lacan, 'The Mirror Stage', in *Jacques Lacan, Écrits* (Paris: Éditions du Seuil, 1966).

11 Artist's website and documentation, <http://www.paulsermon.org/playground/>.

12 Ibid.

within an interactive experience of a collective memory of the playground. The audio sequences were recorded using binaural microphones which placed the sounds spatially as they were when recorded. Additional visual references to this augmented narrative were provided as video clips projected in the interior of the tunnel of the children's train ride. A combination of slow-motion and strobe-flashing image sequences took the visitor further into this augmented memoryscape, a momentary return to the history and collective memory of the environment.

Augmented reality involves the overlaying of digital information onto real space. By moving through the real environment users experienced the digital information at the location to which it referred. Headphones were connected to a Hewlett Packard iPAQ PDA, a small hand-held computer, that played the appropriate sound file corresponding to the position of the user in the playground. The location of the user was determined by a GPS (Global Positioning System) receiver unit attached to the PDA. The playing of sound and video sequences was defined according to a software authoring tool. The authoring tool used a map of the area as a background on to which regions were drawn. Specific commands were associated with each region and defined what the user should experience when they entered or re-entered the space, and a client program running on the PDA worked out which sound file should be played depending on where they were in the region.[13]

## Conclusion

I have tried to keep as much as possible of the original notes, video recordings, and documentation of all my projects. I started doing this with with *Telematic Dreaming* in 1992, even though then I had no idea how important this material would become. It is interesting that I would hear my own commentary on a project – often no more than a couple of sentences, or a kind of short story – told by others, who often had not seen the installation, and told with perfect clarity.

Each time *Telematic Dreaming* is shown it is as if the empty bed is filled with potential. When the 'story' that allows the arrangement and idea of the piece to be understood is told, it enables an immediate understanding of that potential. The participant or user becomes the creator or artist, and my role is to imagine and make possible the space for all that is potential to happen. For me the archived 'lineout' images are the single most important archived element of these telematic encounters, such as that shown in Figure 9.1. They are far more significant than documented images of the installation, representing as they do the moment of communication, meaning and creation of the work.

I have a considerable archive of this recorded lineout video feed. Watching it is something like a portal to the past; you are looking at the very images that

13 Software development by HP Labs, Bristol. Supported by the University of Salford, UK, with financial assistance from Arts Council England and the British Council, Taipei.

caused the scene or effect you are looking at. As the viewer of this archive watches this continual loop of self-reflective experience, the emotions and sensations are brought back to life. There is something almost spiritual about this.

What remains for me as the developer or composer of these installations is the concept – a few pieces of paper or a set of instructions as simple as: 'projector mounted to ceiling, monitors positioned left and right, camera connected to mixer etc.' Like a musical score the concept remains the same, even as the instruments upon which and circumstances in which it is played change.

# Chapter 10
# Tools, Methods, Practice, Process … and Curation

Beryl Graham

When you're holding a hammer, everything looks like a nail.

Jon Winet[1]

Tools are important things, and can affect the way in which people see things. As artist Jon Winet found when working at Xerox Parc research centre: after a while, everything started to look like 'a document', and he found himself considering the pithy American folk saying about hammers. To extend the context, those working in art research might well find themselves clutching inherited methodological tools which, although strong, may not be suitable for the task in front of them.

The examples in this volume give insights into the real-world applications of finding methodological tools that are suited to researching digital media practice. These tools often themselves use digital media, which adds another layer of complexity. As each of the different kinds of digital tools has its own history of hyperbole, the artists' ability to take every ICT tool with a pinch of practical salt is to be admired.

This chapter attempts to put these examples in the wider context of artists who use different parts of the wide spectrum of digital media – in particular, artists using databases and software. They use these media playfully yet revealingly, for art rather than formal research, although the tools themselves may be familiar to arts and humanities researchers. It also explores how changing the methodological tools might most radically change the processes of research. In attempting this, it should be declared that the tools that I clutch are a strange collection of Burginesque media theory, art and design practice research methods, plus the tools of curating new media art.[2] The methods of curating – finding, arranging,

1 J. Winet, 'Riding the Tiger' (1999), <http://www.uiowa.edu/~interart/jwinet/dossier/works/riding_the_tiger.pdf>. (All URLs current at the time of writing.)

2 Victor Burgin's media theory informed my undergraduate education, but much of this chapter is informed by art practice research at the University of Sunderland, and research in my subject area of curating New Media art <http://www.crumbweb.org>. I use the term New Media art to denote digital media and computer-based art which uses the characteristics of interactivity, connectivity and computability in particular, as cited later in this chapter, and explained in B. Graham, 'Redefining Digital Art: Disrupting Borders',

interpreting and disseminating – have also been radically transformed by the new media of networks and participative media, and this chapter also touches on the potential of curatorial tools, whilst bearing in mind the danger that, of course, everything might look like a potential curatorial project!

This vision of a clanking tool belt might raise the suspicion that being a 'jack of all trades' means that there is mastery of none. This 'anxiety of interdisciplinarity'[3] entails the risk that neither discipline will be fully satisfied, and several of the works in this volume speak of the 'tension' between fields. A key tension is that between the languages and methods of art and science, necessitating the kinds of translation of vocabularies spoken of by Jane Prophet. The different values of science and art in relation to the literal translations of scientific visualizations are discussed by Elaine Shemilt. Stephen Scrivener also describes the tensions between fine art and research, in particular relating to the value system of the art market, and usefully categorizes four perspectives on artistic research. The related tension between art and design methods introduced here by Scrivener is taken further by several writers, including the value system of utilitarianism as touched upon by Stelarc. Even within art itself, Janis Jefferies highlights the important differences between an art theory approach to research, and art practice methods which challenge the dominance of text. However, those with experience of overcoming accepted hierarchies of value systems across disciplines can offer methods that might help resolve the validity of knowledge: Janis Jefferies, who has first-hand experience of art/craft hierarchies of knowledge relating to feminist art in the 1980s, interestingly uses the term 'consciousness raising' to describe how making the newer art-practice methods visible helps gradually to change the value systems of conventional methodologies.

These tensions have also been explored elsewhere, but what this volume offers are some steps towards resolutions to be found in the increasingly specific vocabularies used to categorize the function of methods of production and research. Jane Prophet, for example, is very clear on the different roles of the mind map or the maquette and the long time-scale needed for effective interdisciplinary work. Tool belts with too many tools can weigh a researcher down – the need to be an expert in each methodology can leave some dissertations struggling under lengthy methodological chapters when those from a more established field can skip along after a brief standard paragraph. Add to this the need to master different digital media, which might range from biotechnology to computer programming, with what Stelarc admits may be 'limited expertise', and the burdens of ICT methods for new media practice research are considerable. Even something as well-established as using email for discussing art projects at a distance, such as those described by

in F. Cameron and S. Kenderdine (eds), *Theorizing Digital Cultural Heritage: A Critical Discourse* (Cambridge, MA: MIT, 2007), pp. 93–112.

3 A. Coles and A. Defert (eds), 'The Anxiety of Interdisciplinarity', from the series *de-, dis-, ex-. Interdisciplinarity: Art|Architecture|Theory*, vol. 2 (n.p.: BACKless Books/Black Dog, 1998).

Paul Sermon, presents a method that could be regarded as relatively new. Even something as technologically simple as an academic discussion list, such as the CRUMB list quoted here, presents a wide range of methods when considered as a research tool.[4]

However, the fascinating potential of digital media tools is what keeps the explorations going. There are certain themes which differentiate digital media from other media when considered as research tools, and in the spirit of using clear vocabularies and categories it is useful to consider these differentiations rather than the hyperbole of 'the new'.

## Art tools or educational tools?

Although digital media can carry many different kinds of content, and the tools can be used for many purposes, some forms are dominant: walk into most art galleries or museums, and you can justifiably assume that the computers, digital headphones and audio guides carry education about art, rather than art.[5] Screen-based works use interaction to inform via engagement, museum web pages distribute widely 24 hours a day, and databases of digitized art collections or archives can be searched, manipulated and tagged. These desirable characteristics or behaviours of digital media have been categorized as Interactivity, Connectivity and Computability by Steve Dietz,[6] and as he uses them to describe net art, they are characteristics clearly desired by artists as well as educators.

Artists will often test new media to destruction, but will learn important things through that testing. Artists may even satirize the media itself, and attempt to get beyond the hyperbole of speed, power, logic and interaction that sometimes characterizes computer-based media. The collections database, a tool that will be familiar to researchers across a wide range of disciplines, has come under particular scrutiny from artists. *The Unreliable Archivist* (1998) by Janet Cohen, Keith Frank and Jon Ippolito questions not only the point of interaction, but also the process of digital archiving itself. It works like a collections database, and, in true educational interactive style, the user has choices to make and can

4 This is discussed in B. Graham and V. Gfader, 'Curator as Editor, Translator or God? Edited CRUMB Discussion List Theme', *Vagueterrain*, 11 (2008), <http://vagueterrain.net/journal11/crumb/01>.

5 This particular confusion between educational new media and art new media is explored in B. Graham and S. Cook, 'A Curatorial Resource for Upstart Media Bliss', in D. Bearman and J. Trant (eds), *Museums and the Web 2001: Selected Papers from an International Conference* (Pittsburgh: Archives and Museum Informatics, 2001), <http://www.archimuse.com/mw2001/papers/graham/graham.html>, pp. 197–208; and B. Graham and S. Cook, *Rethinking Curating: Art After New Media* (Cambridge, MA: MIT Press, 2010).

6 S. Dietz, 'Why Have There Been No Great Net Artists?', in *Through the Looking Glass: Critical Texts* (1999), <http://www.voyd.com/ttlg/textual/dietz.htm>.

choose the language, images, style and layout. In this case however, the techie-looking slide bars may end up choosing language that is 'loaded', images that are 'enigmatic' and a style that is 'preposterous'. The images on screen are taken from the äda'web collection of new media art, but rather than being treated as sacred museum objects, they are shamelessly manipulated in a way that digital media does very well. Beneath the witty visuals of this piece lie some very serious points which relate to Ippolito's research, as an artist and curator, on the archiving and preservation of 'variable media'.[7] As Paul Sermon touches on in his contribution to this volume, how does one archive the sheer variability and dynamism of artwork with interactive, connective and computable behaviours? Is a collection still a collection if it changes? In relation to databases as an interactive research tool, *The Unreliable Archivist* points out the very circumscribed nature of the choices that the user can make, and the often meaningless interactive features.

More recently, in 2004 the *Database Imaginary*[8] exhibition curated by Sarah Cook, Steve Dietz and Anthony Kiendl examined databased art from Muntadas's *The File Room* (1994–2004), which collected evidence of artistic censorship around the world in analogue filing cabinets and, more recently, online too. Alan Currall's *Encyclopaedia* (2000) is another collectively authored database – this time a collection of videos of homespun definitions of the kind of things that might be defined in encyclopaedia entries, but expressed as talking heads in domestic sitting rooms. Currall's video strategy in particular, in challenging the authority of information and knowledge, could be seen as a precursor to the more recent growth in the use of Wikis, where anyone may add to *or edit existing records* comprising online databases of information. This ability to edit marks Wikis out as behaving differently from databases (which might simply collect together different records from those submitting from various places), and wikis can quickly become more than the sum of their parts, as in the case of Wikipedia. Despite much debate on the accuracy of a minority of individual definitions, Wikipedia has reached a level of authority, and a particular ICT production methodology which promises much for research as well as art. In the *Database Imaginary* exhibition, this was reflected by the work of the *Faculty of Taxonomy* (2004) by University of Openess [sic], where the complex subjects of categories and taxonomies of new media were discussed and developed collaboratively, using a wiki as a development tool, and tactics such as playing games of categorization.

The importance of categorizing and taxonomizing is well-established in research and research methodology. In this volume, Scrivener illuminates by his naming of four categories of approaches to art practice as research, and Paul

7 A. Depocas, J. Ippolito and C. Jones (eds), *Permanence Through Change: The Variable Media Approach* (New York: Guggenheim Museum, 2003), <http://www.variablemedia.net/e/preserving/html/var_pub_index.html>.

8 S. Cook, S. Dietz and A. Kiendl, *Database Imaginary* (exhibition) 21 November–23 January (Banff, Alberta: Walter Phillips Gallery, Banff Center for the Arts, 2004–), <http://databaseimaginary.banff.org>.

Brown navigates the complexities of different concepts of artificial intelligence by the enjoyable and comprehensible 'GOFAI – Good Old Fashioned Artificial Intelligence' et al. Categorizing and taxonomizing have traditionally been the role of the expert, or the curator. Some collections databases can only be searched using keywords defined by the originators of the database, and as *The Unreliable Archivist* illustrates, the very circumscribed nature of the keywords can limit the usefulness of the tool, especially when considering 'the new'. What new media can offer, however, is the ability for the user to develop their own keywords. A hint at this potential can be seen in databases where it is possible to see 'popular searches' to see how other people have been searching, or when we tag, save bookmarks in folders and start to arrange our information. However, the full potential of the media is where the keyword categories and searches of the users can collectively affect the taxonomy of how the collection is arranged – a 'folksonomy'. An example of this is the *runme* (2003–) website. This is a collection of software art where artists may submit and keyword/categorize their own artworks.[9] So far it is a collective database, with tagging. The folksonomy aspect is where those keywords, together with the keywords of those searching the website, go together to make a 'keyword cloud' – those words used repeatedly become firmer categories for arranging the artworks, those poorly used fall out of use. The website programmers still do a certain amount of arranging, so a traditional more expert taxonomy runs alongside the folksonomy, but it seems sensible that as there are few established categories of different kinds of software art, the artists and users help define those categories. The *runme* website is run by a group including software artist Alexei Shulgin, and here the boundaries between *runme* as an educational tool and an art tool are very blurred. It has been described as new media art, and certainly uses new media art sensibilities in the choice of folksonomies as a political position related to open-source methodologies (about which more later). The boundaries are also blurred here between *runme* as an art tool and curatorial tool, as arranging and taxonomizing can also be seen as something that curators do. This ability of digital media – which is after all a set of very diverse kinds of media with very different behaviours and functions – to cross the lines between categories, is also typical of another boundary explored by works in this volume, that between concept or tool.

## Concept or tool?

> 'Why' tends to be left out of most art dialogs I've come across where the why is usually brushed off as an artist 'obsession' or just using whatever tools were available. [...] In the academic/research side of things, the why is part of the work from its inception and being able to tightly knit this idea with the work

9 <http://www.runme.org>.

> ultimately becomes central to the work itself, instead of an afterthought or mere justification. (Jonah Brucker-Cohen)[10]

In a discussion concerning the relationship between formal research and curating new media art, artist and researcher Jonah Brucker-Cohen describes the process and the product, the why of concept and tools, as running throughout the research and artwork, rather than an afterthought. Likewise, in this volume, Gordana Novakovic describes the concept and the tools as 'inseparable'. An integrated process in terms of concept, tool, method and time seems to be important here – as Scrivener also states in this volume, an estimated 20 per cent of time is spent doing research, but this is actually very difficult to separate from time spent teaching, making art, exploring method or other art activities. It could be argued, again, that new media works across the boundaries between method and meaning – the Turing Test as described by Paul Brown in this volume, for example, is a scientific method for defining the existence of artificial intelligence, but it is a method inseparable from Turing's life experience, and from cultural meaning, which continues to inspire artists in very different ways.

When Steve Dietz, as mentioned in this chapter, described categories of net art in terms of Interactivity, Connectivity and Computability, he was describing them in terms of *what they do*, rather than what might traditionally appear on the wall label for artwork, i.e. the materials or tools, be that PHP software or servers running Apache. The media tools, of course, may afford different behaviours, but as each might afford several behaviours it becomes most important to curators to be able to describe how the user experiences the behaviours and the artists' concept, which are inseparable. It is not only artists, but also curators, who are affected by these behaviours: when curator Christiane Paul discussed introducing new media art to her contemporary art colleagues, she found that it was not so much the new tools and media which disturbed them, but the new ways in which she was working – what she did, her process, or method.[11]

Being open about the process and method as well as the tools is therefore an important potential part of digital media. For digital media, openness is not only a choice but a particular methodology built into certain concepts of production. Open source production methods are a particular system of collective software production where the source code of the software is open to others and how it works is transparent to other programmers so that they can edit and adapt what

10 J. Brucker-Cohen, 'Formal Research: Response', *New-Media-Curating Discussion List* (2003), <http://www.jiscmail.ac.uk/lists/new-media-curating.html>. This discussion theme is also available as an edited text file at <http://www.crumbweb.org/discussionArchive.php??>.

11 In particular, Paul mentions collaborative curating methods. B. Graham, 'Edits from a CRUMB Discussion List Theme', in J. Krysa (ed.), *Curating Immateriality: The Work of the Curator in the Age of Network Systems*, Data Browser 03 (London: Autonomedia, 2006), pp. 209–20.

the code does. If wikis can allow the users to change the content of the work, and folksonomies can change how the work is arranged, then open source methods can change the software, the system, and the concept. Although open source as a method does not always translate completely to fields other than software production,[12] researchers, artists, educators and curators are also keen to change the system, including definitions of what research is, and changing systems at a policy level. Jon Ippolito, for example, asks 'how can academics nourish the ecosystem for new media research?', and recommends publishing widely, including via research networks such as Thoughtmesh, renegotiating copyright, and changing the very criteria for research and new media, beyond the traditional journal ranking.[13]

## Which tools?

The complexity of choices and the problems of dividing process from tool from method certainly add some challenges but also offer some particular opportunities for users of ICTs, as long as this does not lead to an unwieldy, clanking tool belt.

### *Building on existing skills*

Artists using digital media usually have a good understanding of the behaviours afforded by those various media, be they interaction or computer systems modelling. A lot of what is new about New Media is that it works across time and space in a different way. Yes, computer communication and processing can be fast, but, as Jane Prophet explains here, the actual process can be a slow and painstaking one. It is true that even something as simple as email can enable research across physical distances, but, as Paul Sermon's work explains, that does not replace face-to-face interaction between people. This intimate understanding of the affordances can help in resisting the hype around both digital media and new methods, and can also help to identify realistic tools, even if those tools are quite low technology:

> For me, that 'fiddling' is often about developing a model as a tool for thinking, so it might be concept diagrams, or I might think of a shape and cut it or tear it out of paper, and fold it as I think about it. Most of those things that get thrown away are invisible, it is so intuitive it is almost invisible work. (Jane Prophet)

---

12 M. Vishmidt with M.A. Francis, J. Walsh, and L. Sykes, *Media Mutandis: A NODE. London Reader. Surveying Art, Technologies and Politics* (London: NODE/Mute, 2006).

13 ThoughtMesh <http://thoughtmesh.net>. J. Ippolito, 'Re: [NEW-MEDIA-CURATING] Exclusivity and Heresy | Alternative academic criteria', *New-Media-Curating Discussion List* (4 May 2008), <http://www.jiscmail.ac.uk/lists/new-media-curating.html>.

Because artists work across the conceptual and the physical, using whatever tools are necessary, they are good at adapting tools to use. Because they have to translate complex concepts such as artificial intelligence across the disciplines of technology and art, artists can become adept at translating the vocabularies and establishing useful working definitions.

What curators and artists tend to be good at is using their skills at visual categorization and arranging objects to inform the communication of research with less dependence upon text. Curator Chris Dorsett, for example, inspired by the 'taxonomies of tasting' used by scientists in the Amazon to identify plant species which traditional visual taxonomies cannot deal with, has suggested that artists' and curator's visual taxonomies of 'arranging things' can lead to useful methods.[14]

### *Working across disciplines – collaboration?*

The anxiety of interdisciplinarity is a key issue for this volume, whether between art and science as outlined in Charlie Gere's introduction, between art and research, or between art and design. The anxiety is compounded by the fact that fields such as 'interaction design' are in themselves already a hybrid of knowledge from design, human–computer interaction and psychology.[15] To overcome this anxiety, collaboration is, as Janis Jefferies states, often very important for universities, but collaboration takes time and understanding.

Even between the relatively close neighbours of art and design, the schisms are often found in the method. To put it simply, design inherits the basic production method of the design cycle of brainstorming, prototyping, feedback and improvement, which enables the rejection at an early stage of things that are a dead end. Art does not inherit this, but is perhaps, as with the artworks mentioned in this chapter, more likely to question received wisdom and to accept knowledge from the powerless as well as the powerful. As I put it in an article, design murders its children and art murders its parents, and each could learn a little from the other.[16] Although many of the writers in this volume are artists, Novakovic describes her artwork as 'cycles', Brown admits that certain tools for art were a 'cul-de-sac' and Prophet says that 'a lot of the results of that activity get thrown away rather than kept'. This could indicate that methods from design are greatly informing art practice even where no collaboration across disciplines is intended.

14 C. Dorsett, 'Exhibitions and their Prerequisites', in J. Rugg (ed.), *Issues in Curating, Contemporary Art and Performance* (London: Intellect, 2007).

15 C. Schubiger, 'Interaction Design: Definition and Tasks', in G. Buurman (ed.), *Total Interaction: Theory and Practice of a New Paradigm for the Design Disciplines* (Basle: Birkauser, 2005), pp. 340–51.

16 B. Graham, 'What Could Art Learn from Design, What Might Design Learn from Art?', in K. Friedman and D. Durling (eds), Proceedings of the Conference *Doctoral Education in Design: Foundations for the Future* (Stoke-on-Trent: Staffordshire University Press, 2000), pp. 425–34.

Collaboration across disciplines is therefore to be encouraged, so that expertise across particular methods can be shared, and perhaps transferred, provided that not too many disciplines are involved. The practicalities of collaborative art and research projects are starting to be made public and well documented, such as the Banff Centre for the Arts New Media Institute events on bridges between research disciplines and collaborative research.[17] The more radical implications of ICT tools, including wikis and folksonomies, where large numbers of people, including users rather than researchers, could be argued to be involved, have yet to be fully rationalized as research methods.

### *Making it public*

The documenting and making public of research, such as the archiving concerns mentioned by Sermon in this volume, are an obvious source of knowledge whereby art research methods can be understood and gain precedent. The descriptions of process and methods as found in this volume are remarkably rare in art writing, and especially in curatorial writing, which tends to concentrate on theory or finished exhibition outputs. An exception to this has been more recent curatorial practice doctoral research. For example, Sarah Cook's 2004 thesis examines the curatorial process in some depth, analysing the constraints and outcomes of various curatorial projects, and naming methods of curating which might address the challenges of particular new media art practices.[18] Joasia Krysa's 2008 thesis also concerns the curation of new media art, including the production of software tools for sharing and analysing software art, translating code into more common language.[19] Because they are open source, these tools can be adapted for use by other curators and programmers, and, in general, if the research and methods are generalizable enough to be usefully adapted by others then that is usefully shared research.

In the dissemination of research, the difference between exposition and exhibition is being debated in relation to equivalency or otherwise to a dissertation,[20]

17 *Bridges Consortium II – New Media Symposium. Searching for New Metaphors, New Practices*, 4–6 October 2002, 'Collaborative Art: Process and Product'. *Summit: Participate/Collaborate: Reciprocity, Design and Social Networks*, 30 September–3 October 2004 (Banff: Banff Centre for the Arts, Banff New Media Institute).

18 S. Cook, 'The Search for a Third Way of Curating New Media Art: Balancing Content and Context in and out of the Institution', PhD thesis, University of Sunderland, 2004.

19 J. Krysa, 'Software Curating: The Politics of Curating in/as (an) Open System(s)', PhD thesis, University of Plymouth, 2008.

20 C. Gray and J. Malins, *Visualizing Research: A Guide to the Research Process in Art and Design* (Oxford: Ashgate, 2004). L. Lyons, 'Walls Are Not My Friends: Issues Surrounding the Dissemination of Practice-led Research within Appropriate and Relevant Contexts', *Working Papers in Art and Design*, 4 (2006), <http://www.herts.ac.uk/artdes/research/papers/wpades/vol4/llfull.html>.

but again, new media tools can blur the boundaries between production and dissemination, with online discussion lists, for example, developing both debate and ideas as they go, developing dialogue and distributing ideas and current research.

### *A good enough method*

Methods using ICT tools can be very powerful, but they only need to be powerful enough to answer a particular research question. Many of the methods in this volume are elegantly simple. Sometimes, there is no need to re-invent a method when an off-the-shelf solution is available. Sometimes, all you need is a hammer.

# Bibliography


Amari, S., 'Theory of Adaptive Pattern Classifiers', *IEEE Transactions in Electronic Computers*, 16 (1967): 299–307.

Ascott, R., *Telematic Embrace* (Berkeley: University of California Press, 2003).

Ashby, W.R., 'Adaptiveness and Equilibrium', *Journal of Mental Science*, 86 (1940): 478–483.

Ashby, W.R., *Design for a Brain* (London: Chapman and Hall, 1952).

Ashby, W.R., *Introduction to Cybernetics* (London: Chapman and Hall, 1956).

Banister, M., *Practical Lithography Printmaking* (New York: Dover, 1972).

Becker, H.S., *Artworlds* (Berkeley and Los Angeles, CA: University of California Press, 1982).

Benjamin, W., *The Work of Art in the Age of Its Technological Reproducibility, and Other Writings on Media* (ed.), M.W. Jennings, B. Doherty and T.Y. Levin (Cambridge, MA: Harvard University Press, 2008).

Bentkowska-Kafel, A., Cashen, T. and Gardiner, H. (eds), *Digital Visual Culture: Theory and Practice* (Bristol: Intellect, 2009).

Bentley, P.J., Novakovic, G. and Ruto, A., '*Fugue*: an Interactive Immersive Audiovisualisation and Artwork using an Artificial Immune System', in *Proceedings of ICARIS 2005, Artificial Immune Systems* (Berlin/Heidelberg: Springer, 2005), 1–12.

Biederman, C., *Art as the Evolution of Visual Knowledge* (Red Wing: MI, 1948).

Biggs, S., *Introduction for ITEM: Artists in Research Environments* <http://hosted.simonbiggs.easynet.co.uk/texts/item.html>, originally published for Liverpool: FACT, 2006.

Bird, J. and Stokes, D., 'Evolving Fractal Drawings', in C. Soddu (ed.), *Generative Art 2006 Proceedings of the 9th International Conference* (2006), pp. 317–27.

Bird, J., Stokes, D., Husbands, P., Brown, P. and Bigge, B., 'Towards Autonomous Artworks', *Leonardo Electronic Almanac* (Cambridge, MA: MIT Press, forthcoming).

Blackwell, T. and Jefferies, J., 'Swarm Tech-tiles', in E. Rothlauf et al. (eds), *Applications of Evolutionary Computing* (Evoworkshops, 2005), pp. 468–77.

Blackwell, T. and Jefferies, J., 'Collaboration: A Personal Report', *International Journal of Co-Creation in Design and the Arts*, 2/4 (2006): 259–63.

Boden, Margaret A., *Mind as Machine: A History of Cognitive Science* (Oxford: Oxford University Press, 2006).

Brighton, C.R., 'Research in Fine Art: An Epistemological and Empirical Study', Sussex University, unpublished PhD thesis (1992).

Brooks, R.A., 'A Robust Layered Control System for a Mobile Robot', *IEEE Journal of Robotics and Automation*, 2/1 (1986): 14–23.

Brooks, R.A., *Cambrian Intelligence: The Early History of the New AI* (Cambridge, MA: MIT Press, 1999).

Brown, P., 'The Mechanisation of Art', in P. Husbands, O. Holland and M. Wheeler (eds), *The Mechanical Mind in History* (Cambridge, MA: MIT Press, 2008), 275–289.

Brown, P., 'From System Art to Artificial Life', in C. Gere, P. Brown, N. Lambert and C. Mason (eds), *White Heat Cold Logic: British Computer Art 1960–1980* (Cambridge, MA: MIT Press, forthcoming).

Brucker-Cohen, J., 'Formal Research: Response', *New-Media-Curating Discussion List* (2003), <http://www.jiscmail.ac.uk/lists/new-media-curating.html>.

Burgin, V., *The End of Art Theory: Criticism and Postmodernity* (London: Palgrave Macmillan, 1986).

Candlin, F., 'Artwork and the Boundaries of Academia: A Theoretical/Practical Negotiation of Contemporary Art Practice within the Conventions of Academic Research', University of Keele, unpublished PhD thesis (1998).

Candlin, F., *Working Papers in Art and Design*, vol. 1 (2000), <http://www.herts.ac.uk/artdes1/research/papers/wpades/vol1/candlin2full.html>.

Candy, L., 'Practice Based Research: A Guide', Creativity and Cognition Studios, University of Technology, Sydney, CCS Report: 2006, V1.0 (2006), <http://www.creativityandcognition.com>.

Candy, L. and Edmonds, E., *Exploration in Arts and Technology* (London: Springer, 2002).

Candy, L. and Edmonds, E., 'Interaction in Art and Technology', *Crossings: eJournal of Art and Technology*, 2/1 (2002), <http://crossings.tcd.ie/issues/2.1/Candy/>.

Chaplin, E., *Sociology and Visual Representation* (London: Routledge, 1994).

Clements, W., 'Surveillance and the Art of Software Maintenance: Remarks on logo_wiki', in *Observatori 2008. After the Future* (Valencia: Observatori.com, 2008).

Cliff, D.T., Harvey, I. and Husbands, P., 'Explorations in Evolutionary Robotics', *Adaptive Behaviour*, 2/1 (1993): 71–108.

Cohen, H., 'Reconfiguring', in P. Brown, C. Gere, N. Lambert and C. Mason (eds), *White Heat Cold Logic: British Computer Art 1960–1980* (Cambridge, MA: MIT Press, 2008), 141–150.

Coles, A. and Defert, A. (eds), 'The Anxiety of Interdisciplinarity', from the series *de-, dis-, ex-. Interdisciplinarity: Art|Architecture|Theory*, vol. 2 (London: BACKless Books/Black Dog, 1998).

Connolly, M., *Art Practice, Peer-Review and the Audience for Academic Research*, Position Papers on Practice-Based Research, National College of Art and Design, Dublin, Ireland, 22 April 2005.

Cook, S., 'The Search for a Third Way of Curating New Media Art: Balancing Content and Context in and out of the Institution', PhD thesis, University of Sunderland, 2004.

Cook, S. and Graham, B., *Curating Art after New Media* (Cambridge, MA: MIT Press, 2010).

Cook, S., Dietz, S. and Kiendl, A., *Database Imaginary* [exhibition] 21 November–23 January (Banff, Alberta: Walter Phillips Gallery, Banff Center for the Arts, 2004–), <http://databaseimaginary.banff.org>.

Cox, D., 'Renaissance Teams and Scientific Visualization: A Convergence of Art and Science', *Collaboration in Computer Graphics Education: Proceedings SIGGRAPH 88 Educator's Workshop* (1988), pp. 81–104.

Craik, K.J.W., *The Nature of Explanation* (Cambridge: Cambridge University Press, 1943).

Dawkins, R., 'Viruses of the Mind', <http://www.cscs.umich.edu/~crshalizi/Dawkins/viruses-of-the-mind.html>.

Depocas, A., Ippolito, J. and Jones, C. (eds), *Permanence Through Change: The Variable Media Approach* (New York: Guggenheim Museum, 2003), <http://www.variablemedia.net/e/preserving/html/var_pub_index.html>.

Dietz, S., 'Why Have There Been No Great Net Artists?', in *Through the Looking Glass: Critical Texts* (1999), <http://www.voyd.com/ttlg/textual/dietz.htm>.

Doidge, N., *The Brain That Changes Itself* (New York: Viking, 2007).

Dorsett, C., 'Exhibitions and Their Prerequisites', in J. Rugg (ed.), *Issues in Curating, Contemporary Art and Performance* (London: Intellect Press, 2007), pp. 77–87.

Durling, D., Freidman, K. and Gutherson, P., 'Editorial: Debating the Practice-Based PhD', *International Journal of Design Sciences and Technology*, 10/2 (2002): 7–18.

Edelman, G., *Bright Air, Brilliant Fire: On the Matter of the Mind* (New York: Basic Books, 1993).

Edmonds, E.A. et al., 'The Studio as Laboratory: Combining Creative Practice and Digital Technology Research', *International Journal of Human-Computer Studies*, 63/4–5 (2005): 452–81.

Ehrenzweig, A., *The Hidden Order in Art. A Study in the Psychology of Artistic Imagination* (Berkeley: University of California Press, 1967).

Floreano, D. and Mondada, F., 'Automatic Creation of an Autonomous Agent: Genetic Evolution of a Neural-Network Driven Robot', in D. Cliff, P. Husbands, J. Meyer and S.W. Wilson (eds), *From Animals to Animats III: Proceedings of the Third International Conference on Simulation of Adaptive Behavior* (Cambridge, MA: MIT Press–Bradford Books, 1994), pp. 402–10.

Floreano, D., Husbands, P. and Nolfi, S., 'Evolutionary Robotics', in B. Siciliano and O. Khatib (eds), *Springer Handbook of Robotics* (Berlin: Springer, 2008), pp. 1423–51.

Foster, H., *The Return of the Real* (Cambridge, MA: MIT Press, 1996).

Fourmetraux, J.-P., 'Governing Artistic Innovation', *Leonardo*, 40/5 (2007), 489–492.

Frayling, C., 'Research in Art and Design', Royal College of Art Research Papers, 1/1 (1993/4): 1–5.

Frayling, C., 'Foreword', in K. MacLeod and L. Holdridge, *Thinking through Art: Reflections on Art as Research* (London, Routledge, 2006), pp. xiii–xiv.

Gere, C., 'False Divide Between Art and Science', *The Guardian*, Saturday 4 August 2007.

Graham, B., 'What Could Art Learn from Design, What Might Design Learn from Art?', in K. Friedman and D. Durling (eds), Proceedings of the Conference, *Doctoral Education in Design: Foundations for the Future* (Stoke-on-Trent: Staffordshire University Press, 2000), pp. 425–34.

Graham, B., 'Edits from a CRUMB Discussion List Theme', in J. Krysa (ed.), *Curating Immateriality: The Work of the Curator in the Age of Network Systems*, Data Browser 03 (London: Autonomedia, 2006), pp. 209–20.

Graham, B., 'Redefining Digital Art: Disrupting Borders', in F. Cameron and S. Kenderdine (eds), *Theorizing Digital Cultural Heritage: A Critical Discourse* (Cambridge, MA: MIT, 2007), 93–113.

Graham, B. and Cook, S., 'A Curatorial Resource for Upstart Media Bliss', in D. Bearman and J. Trant (eds), *Museums and the Web 2001: Selected Papers from an International Conference* (Pittsburgh: Archives and Museum Informatics, 2001), <http://www.archimuse.com/mw2001/papers/graham/graham.html>, pp. 197–208.

Graham, B. and Gfader, V., 'Curator as Editor, Translator or God? Edited CRUMB Discussion List Theme', *Vagueterrain*, 11 (2008), <http://vagueterrain.net/journal11/crumb/01>.

Gray, C., 'Inquiry through Practice: Developing Appropriate Research Strategies in Art and Design', in P. Strandman (ed.), *No Guru, No Method* (Helsinki: University of Arts and Design, 1998), pp. 82–9.

Gray, C. and Malins, J., *Visualizing Research: A Guide to the Research Process in Art and Design* (Aldershot: Ashgate, 2004).

Grossberg, S., 'Contour Enhancement, Short-Term Memory, and Constancies in Reverberating Neural Networks', *Studies in Applied Mathematics*, 52 (1973): 213–57.

Guaridis, A., Infonoise, INFONOISE interactive gallery installation and web-connected theatre event (website; updated 2002), <http://www.infonoise.net/news/Angueliki.html>.

Hauser, J. (ed.), *sk-Interfaces. Exploding Borders – Creating Membranes in Art, Technology and Society* (Liverpool: Liverpool University Press, 2008).

Heims, S., *Constructing a Social Science for Postwar America: The Cybernetics Group, 1946–1953* (Cambridge, MA: MIT Press, 1991).

Hin, A.M., *An Introduction to a History of Woodcut* ([New York: Houghton Mifflin, 1935], New York: Dover Publications, 1963).

Hobbes, T., *Leviathan* (London: Andrew Crooke, 1651).

Holland, J.H., *Adaptation in Natural and Artificial Systems* (Ann Arbor: University of Michigan Press, 1975).

Husbands, P. and Harvey, I., 'Evolution versus Design: Controlling Autonomous Mobile Robots', in *Proceedings of the 3rd Annual Conference on Artificial Intelligence, Simulation and Planning in High Autonomy Systems* (Los Alimitos, CA: IEEE Computer Society Press, 1992), pp. 139–46.

Husbands, P. and Holland, O., 'The Ratio Club: A Hub of British Cybernetics', in P. Husbands, O. Holland and M. Wheeler (eds), *The Mechanical Mind in History* (Cambridge, MA: MIT Press), pp. 91–148.

Husbands, P., Holland, O. and Wheeler, M. (eds), *The Mechanical Mind in History* (Cambridge, MA: MIT Press, 2008).

Imperato, A., *A_Imperato_Resume*, <http://alessandroimperato.com/Resume-Contact_files/A_Imperato_Resume.pdf>.

Ippolito, J., 'Re: [NEW-MEDIA-CURATING] Exclusivity and Heresy | Alternative academic criteria', *New-Media-Curating Discussion List* (2008), <http://www.jiscmail.ac.uk/lists/new-media-curating.html>.

Jakobi, N., 'Evolutionary Robotics and the Radical Envelope of Noise Hypothesis', *Adaptive Behaviour*, 6 (1998): 325–68.

Kauffman, L.H. and Varela, F.J., 'Form Dynamics', *Journal of Social Biological Structures*, 3 (1980): 171–206.

Kaye, N., *Site-Specific Art: Performance, Place, Documentation* (London: Routledge, 2000).

Klee, P., *The Diaries of Paul Klee 1898–1918* (ed.), F. Klee (Berkeley, CA: University of California Press, 1964).

Krauth, N., 'The Preface as Exegesis', *TEXT*, 6/1 (2002), 1–15.

Krysa, J., 'Software Curating: The Politics of Curating in/as (an) Open System(s)', PhD thesis, University of Plymouth (2008).

Kuhn, T.S., *The Structure of Scientific Revolutions* (Chicago: University of Chicago Press, 1970).

Lacan, Jacques, 'The Mirror Stage', in *Jacques Lacan, Écrits* (Paris: Éditions du Seuil, 1966).

Latham, W. and Todd, S., *Evolutionary Art and Computers* (London: Academic Press, 1992).

Latour, B., *Science in Action: How to Follow Scientists and Engineers through Society* (Cambridge, MA: Harvard University Press, 1987).

Latour, B. and Woolgar, S., *Laboratory Life: The Social Construction of Scientific Facts* (Beverly Hills: Sage Publications, 1979).

Lippard, L.R., *Six Years: The Dematerialization of the Art Object from 1966 to 1972* (Berkeley: University of California Press, 1973, 1997).

Lovejoy, M., *Postmodern Currents: Art and Artists in the Age of Electronic Media*, 2nd edn (New Jersey: Prentice Hall, 1997).

Lury, C., *Brands: The Logos of the Global Economy* (London: Routledge, 2004).

Lyons, L., 'Walls Are Not My Friends: Issues Surrounding the Dissemination of Practice-Led Research within Appropriate and Relevant Contexts', *Working Papers in Art and Design 4* (2006), <http://www.herts.ac.uk/artdes/research/papers/wpades/vol4/llfull.html>.

McCarthy, J., Minsky, M., Rochester, N. and Shannon, C., 'A Proposal for the Dartmouth Summer Research Project on Artificial Intelligence' (1955), <http://www-formal.stanford.edu/jmc/history/dartmouth/dartmouth.html>.

McCulloch, W.S. and Pitts, W., 'A Logical Calculus of the Ideas Immanent in Nervous Activity', *Bulletin of Mathematical Biophysics*, 5 (1943): 115–33.

McFadden, J., 'A Genetic String Band', *The Guardian*, Friday 3 August 2007.

McLuhan, M., *Understanding Media* (London: Sphere, 1971).

Macleod, K. and Holdridge, L., 'Introduction', in K. Macleod and L. Holdridge, *Thinking through Art: Reflections on Art as Research* (London: Routledge, 2006), pp. 15–19.

Makela, M., 'Knowing Through Making: The Role of the Artefact in Practice-Led Research', *Journal of Knowledge, Technology and Policy*, 20/3 (2007), 157–163.

Manovich, L., *The Language of New Media* (Cambridge, MA: MIT Press, 2001).

Mason, C., 'A Computer in the Art Room' <http://www.chart.ac.uk/chart2004/papers/mason.html> (2004).

Mason, C., *A Computer in the Art Room: The Origins of British Computer Arts 1950–80* (Norfolk: JJG, 2008).

Meggs, P.B., *A History of Graphic Design* (New Jersey: John Wiley and Sons, 1998).

Minsky, M.L. and Papert, S.A., *Perceptrons* (Cambridge, MA: MIT Press, 1969).

Moravec, H., 'The Stanford Cart and The CMU Rover', *Proceedings of the IEEE*, 71/7 (1983): 872–84.

Moravec, H., 'Sensing Versus Inferring in Robot Control', Informal Report (1987), <http://www.frc.ri.cmu.edu/~hpm/project.archive/robot.papers/1987/sense.lt>.

Nilsson, N.J. (ed.), *Shakey the Robot*, Technical Note 323 (Menlo Park, CA: AI Center, SRI International, 1984).

Nolfi, S. and Floreano, D., *Evolutionary Robotics: The Biology, Intelligence, and Technology of Self-Organizing Machines* (Cambridge, MA: MIT Press/Bradford Books, 2000).

Novakovic, G. (ed.), *The Shirt of a Happy Man: Multimedia and Interactive Artworks and Projects 1989–1995* (Belgrade: Soros Foundation, Department for Contemporary Art, 1995).

Novakovic, G., 'INFOWAR: Info Noise in Belgrade – part 1', INFOWAR Ars Electronica 98 Linz Austria netsymposium discussion (website; updated 9 June 1998), <http://www.aec.at/infowar/NETSYMPOSIUM/ARCH-EN/msg00144.html>.

Novakovic, G., 'Shirt' (published online 1998; updated September 2005), <http://www.gordananovakovic.net/work/shirt/index.html> .

Novakovic, G., Works 1998 to 2001 (published online 2001), <http://www.gordananovakovic.net/writings/works1998_2001.html>.

Novakovic, G., 'Electronic Cruelty', in R. Ascott (ed.), *Engineering Nature: Art and Consciousness in the Post-Biological Era* (Bristol: Intellect, 2006).

Novakovic, G., 'Metropolis: An Extreme and Hostile Environment', in *MutaMorphosis: Challenging Arts and Sciences*, Conference Proceedings, <http://mutamorphosis.wordpress.com/2009/01/15/gordana-novakovic/>.

Novakovic, G. and Savić M., *Under the Shirt of a Happy Man* (Belgrade: Soros Foundation, Department for Contemporary Art, 1995).

Novakovic, G., Bentley, P. and Ruto, A., Tesla – Art and Science Research Interest Group (website updated 2008), <http://www.cs.ucl.ac.uk/research/tesla/>.

Novakovic, G., Linz, R. and Milkovic, Z., Infonoise, INFONOISE Interactive Gallery Installation and Web-Connected Theatre Event (website updated 2002), <http://www.infonoise.net>.

Novakovic, G., Linz, R., Bentley, P. and Ruto, A., 'FUGUE Art and Science Collaboration' (website updated 2008), <http://www.fugueart.com>. .

Paul, C., *Digital Arts* (London: Thames and Hudson, 2003).

Perris, M., 'Evolving Ecologically Inspired Drawing Behaviours', MSc dissertation, Dept of Informatics, University of Sussex (2007).

Reichardt, J., 'In the Beginning', in C. Gere, P. Brown, N. Lambert and C. Mason (eds), *White Heat Cold Logic: British Computer Art 1960–1980* (Cambridge, MA: MIT Press, 2008), 71–81.

Rosenblatt, F., 'The Perceptron: A Probabilistic Model for Information Storage and Organization in the Brain', *Cornell Aeronautical Laboratory, Psychological Review*, 65/6 (1958): 386–408.

Rust, C., Mottram, J. and Till, J., *AHRC Research Review: Practice-Led Research in Art, Design and Architecture* (February 2008), <http://www.ahrc.ac.uk/apply/research/sfi/ahrcsi/research_reviews>.

Schaffer, S., 'Fish and Ships: Models in the Age of Reason', in S. de Chadarevian and N. Hopgood (eds), *Models: The Third Dimension of Science* (Stanford: Stanford University Press, 2004), 71–105.

Schubiger, C., 'Interaction Design: Definition and Tasks', in G. Buurman (ed.), *Total Interaction: Theory and Practice of a New Paradigm for the Design Disciplines* (Basle: Birkauser, 2005), 340–51.

Scrivener, S.A.R., 'Change Factors and the Contribution Made by Research to the Disciplines', unpublished report (2007).

Scrivener, S.A.R., 'The Roles of Art and Design Process and Object in Research', in N. Nimkulrat and T. O'Riley (eds), *Reflections and Connections: On the Relationship between Creative Production and Academic Research* [e-book] (Helsinki: Helsinki University of Art and Design, 2009).

Selfridge, O.G., 'Pandemonium: A Paradigm for Learning', in D. Blake and A. Uttley (eds), *The Mechanisation of Thought Processes*. Volume 10 of National Physical Laboratory Symposia (London: HMSO, 1959), 511–29.

Shapin, S. and Schaffer, S., *Leviathan and the Air-Pump: Hobbes, Boyle, and the Experimental Life* (Princeton, NJ: Princeton University Press, 1985).

Spencer Brown, G., *Laws of Form* (London: Allen and Unwin, 1969).

Sullivan, G., *Art Practice as Research: Inquiry in the Visual Arts* (London: Sage Publications, 2005).

Thompson, J., 'Art Education: From Coldstream to QAA', *Critical Quarterly*, 47/1–2 (2005): 215–55.

Turing, A.M., 'Computing Machinery and Intelligence', *Mind*, 59 (1950): 433–60.

UN Security Council 3082nd Meeting Resolution S/RES/757, 30 May 1992, <http://www.nato.int/IFOR/un/u920530a.htm>.

Vishmidt, M. with Francis, M.A., Walsh, J. and Sykes, L., *Media Mutandis: A NODE.London Reader. Surveying Art, Technologies And Politics* (London: NODE/Mute, 2006).

Walter, W.G., 'An Imitation of Life', *Scientific American*, 182/5 (1950): 42–5.

Weber, M., *On Universities: The Power of the State and the Dignity of the Academic Calling* (Chicago: University of Chicago Press, 1976).

Whitelaw, M., 'The Abstract Organism: Towards a Prehistory for A-Life Art', *Leonardo*, 34/4 (2001): 345–8.

Wiener, N., *Cybernetics: Or the Control and Communication in the Animal and the Machine* (Cambridge, MA: MIT Press. 1948).

Wilson, S., 'Myths and Confusions in Thinking about Art/Science/Technology', paper presented at the College Art Association Meetings in NYC (2000), <http://userwww.sfsu.edu/~swilson/papers/wilson.caapaper.html>.

Wilson, S., *Information Arts: Intersections of Art, Science, and Technology* (Cambridge, MA: MIT, 2002).

Wilson, S., *Research as a Cultural Activity*, <http://userwww.sfsu.edu/~swilson/>, (n.d.).

Winet, Jon 'Riding the Tiger' (1999), <http://www.uiowa.edu/~interart/jwinet/dossier/works/riding_the_tiger.pdf>.

Yaneva, A., 'Scaling Up and Down: Extraction Trials in Architectural Design', *Social Studies of Science*, 35/6 (2005): 867–94.

Young, J.O., *Art and Knowledge* (London: Routledge, 2001).

Zivanovic, A., 'The Technologies of Edward Ihnatowicz', in C. Gere, P. Brown, N. Lambert and C. Mason (eds), *White Heat Cold Logic: British Computer Art 1960–1980* (Cambridge, MA: MIT Press, 2008), 95–110.

# Index

Page numbers in *italics* refer to figures and tables.